Springer Theses

Recognizing Outstanding Ph.D. Research

For further volumes:
http://www.springer.com/series/8790

Aims and Scope

The series "Springer Theses" brings together a selection of the very best Ph.D. theses from around the world and across the physical sciences. Nominated and endorsed by two recognized specialists, each published volume has been selected for its scientific excellence and the high impact of its contents for the pertinent field of research. For greater accessibility to non-specialists, the published versions include an extended introduction, as well as a foreword by the student's supervisor explaining the special relevance of the work for the field. As a whole, the series will provide a valuable resource both for newcomers to the research fields described, and for other scientists seeking detailed background information on special questions. Finally, it provides an accredited documentation of the valuable contributions made by today's younger generation of scientists.

Theses are accepted into the series by invited nomination only and must fulfill all of the following criteria

- They must be written in good English.
- The topic should fall within the confines of Chemistry, Physics, Earth Sciences, Engineering and related interdisciplinary fields such as Materials, Nanoscience, Chemical Engineering, Complex Systems and Biophysics.
- The work reported in the thesis must represent a significant scientific advance.
- If the thesis includes previously published material, permission to reproduce this must be gained from the respective copyright holder.
- They must have been examined and passed during the 12 months prior to nomination.
- Each thesis should include a foreword by the supervisor outlining the significance of its content.
- The theses should have a clearly defined structure including an introduction accessible to scientists not expert in that particular field.

Yuki Sato

Space-Time Foliation in Quantum Gravity

Doctoral Thesis accepted by
Nagoya University, Nagoya, Japan

Author
Dr. Yuki Sato
Department of Physics
Nagoya University
Nagoya
Japan

Supervisor
Prof. Tadakatsu Sakai
Department of Physics
Nagoya University
Nagoya
Japan

ISSN 2190-5053 ISSN 2190-5061 (electronic)
ISBN 978-4-431-56225-2 ISBN 978-4-431-54947-5 (eBook)
DOI 10.1007/978-4-431-54947-5
Springer Tokyo Heidelberg New York Dordrecht London

Softcover reprint of the hardcover 1st edition 2014

Printed on acid-free paper

Springer is part of Springer Science+Business Media (www.springer.com)

… I am inclined to believe from this that four-dimensional symmetry is not a fundamental property of the physical world.

—P. A. M. Dirac in his paper, "The Theory of Gravitation in Hamiltonian Form."

Supervisor's Foreword

It has been one of the outstanding problems in theoretical physics to formulate a quantum theory of gravity. A naive quantization had been made of Einstein's theory of gravity, which is well established as a classical theory of gravity with space-time regarded as a dynamical object. However, this procedure turns out to lead to mathematical inconsistency, because the Einstein's theory is not renormalizable. This may imply that for a proper formulation of quantum gravity, we need a new ingredient that is missing in our current understanding of the physical laws. As an example, it is often argued that string theory is one of the most promising candidates of a quantum gravity. In string theory, the most fundamental degree of freedom in nature is regarded as a string rather than a point-particle. Furthermore, a lattice quantum gravity has been studied intensively so far. In this framework, a continuous space-time is approximated with a discretized manifold that is composed of a number of simplices. Dynamical triangulation (DT) is formulated by replacing a path integral over continuous space-times with a sum over all the way of the discretizations. It is found that DT succeeds in defining quantum gravity in lower dimensions but results in pathologies in realistic four-dimensional quantum gravity. In order to resolve this problem, causal dynamical triangulation (CDT) proposes to incorporate space-time causality into DT. The configurations to be summed over in CDT are restricted by requiring them to be consistent with causality. It is then expected to obtain a consistent quantum gravity in four-dimensional space-time.

One of the main topics in this thesis is to investigate some quantum gravity models in two dimensions using CDT. First, this thesis studies a two-dimensional CDT coupled to dimers. A path integral over the configuration space is made by combinatorics. It is important that the thesis found a new multicritical point that describes the CDT model in a continuum limit. Furthermore, it is argued that this formulation enables one to generalize the CDT model coupled to dimers so that it allows creation and annihilation of baby universes. Next, this thesis discusses a two-dimensional CDT with extended interactions by means of a string field theory technique. In this framework, one can formulate a generalized CDT model with extended interactions using creation and annihilation operators of a baby universe. As well, it is shown that equivalent descriptions based on a matrix model exist.

This thesis also discusses classical gravity with an aim to extend Einstein's theory of gravity. A key idea is to allow only a foliation preserving diffeomorphism rather than full diffeomorphism. A typical example is Hořava-Lifshitz, which is supposed to be a UV-finite quantum gravity. This thesis comes up with a new gravity theory called the n-DBI model, whose action contains infinitely many higher derivative terms. It is argued that in this model the cosmological constant is given as an integration constant that emerges after solving an equation of motion. Moreover, it is shown that this model allows an extra degree of freedom of a physical, propagating graviton.

It is worth mentioning that this thesis is unique; in that, it puts a great emphasis on the foliation structure of space-time and its causality as a guiding principle throughout the thesis.

Nagoya, March 2014 Tadakatsu Sakai

Acknowledgments

First of all, I would like to thank my supervisor, Tadakatsu Sakai, for his discussions, encouragement, and patience. Without his support and well-honed guidance, I could not have so much as submitted my thesis. I would like to express my appreciation to my supervisor in the Niels Bohr Institute, Jan Ambjørn. Without his kind acceptance and many fruitful discussions, I never could have reached the point where I am. "Thank you" does not quite cover it.

Through wonderful collaborations, I learned how to overcome tough situations in physics and shared joy and sorrow. Thus, I would like to acknowledge the following individuals as collaborators as well as good friends: Jan Ambjørn, Flávio S. Coelho, Hiroyuki Fuji, Lisa Glaser, Andrzej Görlich, Carlos Herdeiro, Takayuki Hikichi, Kyosuke Hirochi, Shinji Hirano, Masazumi Honda, Taishi Katsuragawa, Stefano Kovacs, Gaurav Narain, João P. Rodrigues, Naoki Sasakura, Hidehiko Shimada, Hiroshi Suenobu, Tomo Tanaka, and Yoshiyuki Watabiki.

I thank Charlotte Kristjansen and Bergfinnur Durhuus for many discussions and much kind encouragement during my stay in Copenhagen. I would like to acknowledge the members of the CDT seminar, Takayuki Hikichi, Yasusada Nambu, and Hiromi Saida, for enjoyable discussions. I am grateful to my good friends at Nagoya University for sharing wonderful times with me, and I would like to express my appreciation to my good friends at the Niels Bohr Institute. Without them, my time in Copenhagen would have been steeped in darkness just like the Danish winter. During my stay in Copenhagen, I got powerful encouragement from Japanese visitors Hidenori Fukaya, Masao Ninomiya, and Katsuhiko Sato; I thank them. I would also like to thank all faculty members at Nagoya University and the Niels Bohr Institute for strong encouragement and warm hospitality. I am especially grateful to my family members Kumiko Sato, Tetsuo Sato, and Yohei Sato. Thanks to their support with love, I could survive my life as a Ph.D. student.

I express my appreciation to Mihoko Kumazawa and Hisako Niko, editorial staff members at Springer Japan, for helping me publish my thesis and for their earnest and kind correspondence.

I would like to dedicate this thesis to my dear grandmother, Kikue Hakozaki, who passed away during the completion of this work.

Lastly, I would like to take this opportunity to say a few words about one of my good friends, Takayuki Hikichi. He was a Ph.D. student at Nagoya University,

Japan. His attitude toward physics was earnest: I often saw him asking questions with passion at regular seminars. He was always constructive and brave: without being afraid of authority he insisted on what he believed to be right. Due to his daily diligent effort, he had almost finished his own work on some computer simulations of 3-dimensional causal dynamical triangulations. I was shocked and very saddened to hear of Takayuki's loss during the publication process of this thesis. I would like to dedicate my thesis to my dear friend Takayuki Hikichi, as well.

Contents

1 Introduction 1
1.1 Overview 1
1.1.1 A Role of Foliation 1
1.2 Quantum Gravity Without Coordinates 4
1.2.1 General Relativity Without Coordinates 4
1.2.2 Causal Dynamical Triangulations 9
1.2.3 Exactly Solvable Models 12
1.3 Geometrodynamics 21
1.3.1 ADM Formalism 21
1.3.2 Hořava-Lifshitz Gravity 26
1.3.3 *n*-DBI Gravity 29
References 34

2 Causal Dynamical Triangulation 37
2.1 Matter-Coupled CDT 37
2.1.1 Causal Random Geometries Coupled to Dimers 37
2.1.2 Field Theory Qrising from CDT Scaling 42
2.1.3 Discussion 43
2.2 Extended Interactions in CDT 44
2.2.1 Generalized CDT 44
2.2.2 Generalized CDT with Extended Interactions 47
2.2.3 Discussion 53
References 54

3 *n*-DBI Gravity 57
3.1 Black Hole Solutions in *n*-DBI Gravity 57
3.1.1 Solutions with Constant $\mathscr{R}$ 57
3.1.2 Spherically Symmetric Solutions 60
3.1.3 Discussion 64

3.2 Scalar Graviton in n-DBI Gravity 65
3.2.1 Nailing Scalar Graviton 65
3.2.2 Identity of Scalar Graviton 68
3.2.3 Potential Pathologies 72
3.2.4 Discussion 77
References 78

4 Summary 81

Appendix A: Notation and Formulae 83

Appendix B: Computations in n-DBI Gravity 85

CV-Yuki Sato 99

List of Published Articles

Parts of this thesis have been published in the following journal articles:

1. F. S. Coelho, C. Herdeiro, S. Hirano and Y. Sato, "Scalar graviton in n-DBI gravity," Phys. Rev. D **86** (2012) 064009.
2. J. Ambjørn, L. Glaser, A. Görlich and Y. Sato, "New multicritical matrix models and multicritical 2d CDT," Phys. Lett. B **712** (2012) 109.
3. C. Herdeiro, S. Hirano and Y. Sato, "n-DBI gravity," Phys. Rev. D **84** (2011) 124048.
4. H. Fuji, Y. Sato and Y. Watabiki, "Causal Dynamical Triangulation with Extended Interactions in 1 + 1 Dimensions," Phys. Lett. B **704** (2011) 582.

Chapter 1
Introduction

1.1 Overview

Lots of unsolved fundamental problems in theoretical physics are closely tied to the lack of a complete quantum theory of gravity. Once the quantum gravity is established, it should provide us transparent answers to simple questions even children ask their parents such as "*What is time?,*" and "*Why is the Universe the way it is?*" However, as the history tells, the quantization of gravity is an issue that defies any attempt at a quick and simple solution. One difficulty in formulating quantum gravity is the weakness of the gravitational force. To observe quantum gravity effects, one needs to approach the Planck scale, 10^{19}GeV, which is far from the energy scale achieved by current collider experiments. Thus, it is hard to use experimental data as hints to construct quantum gravity. The other difficulty is the non-renormalizability of the Einstein-Hilbert action, according to conventional power-counting arguments of the Newton constant. The most fundamental difficulty is related to locality: due to the general covariance, one cannot define any gauge-invariant local observables in general space-time.

Extracting the truth of nature from the small amount of information, theoretical physicists are approaching to the correct answer or going away from it, although every physicist expects the former scenario more or less. In the case of the author, a notion of the space-time foliation seems to be an efficeint concept in quantum gravity. The reason can be found below.

1.1.1 A Role of Foliation

In 1983, Teitelboim proposed a quite impressive work about the path-integral of gravitational theories [1]. In the paper, he insisted that one should choose either the causality or gauge invariance when quantizing gravity via the path-integral: the two notions can not coexist. His argument does not rely on any specific model of

Y. Sato, *Space-Time Foliation in Quantum Gravity*, Springer Theses,
DOI: 10.1007/978-4-431-54947-5_1, © Springer Japan 2014

quantum gravity, but still one can learn important aspects quantum gravity may have. We start with reviewing his idea. We firstly clarify the notion of space-time. As claimed by Wheeler [2], space-time is the *classical* history of codimension-1 spatial geometry. In this sense, the path-integral of gravity is nothing but summing up probable classical histories (paths) of the codimension-1 spatial geometry. One knows from general relativity (GR) that a classical geometry is the pseudo-Riemannian manifold with the causality built-in. Through the sum over space-time histories, quantum-mechanical nature of space-time emerges: in particular, a notion of time disappears. As what is fundamental in the path-integral of a point particle is not the quantum-mechanical amplitude but each classical path of a point particle, each classical history of space-time is an elementary concept in the path-integral of gravity. The issue is how to include the causality in each history. For preparation to answer this, we mention the most important lesson we have learned from GR, i.e., general covariance. If the time derivative of the metric is of first order at most, one can move onto the Hamiltonian formalism; if the action has general covariance, the corresponding Hamiltonian density can be written as a linear combination of first class constraints (see Sect. 1.3)[1]:

$$\mathscr{H} = N\Phi_4 + N_i\Phi_5^i, \tag{1.1}$$

where N and N_i are the lapse and shift functions; Φ_4 and Φ_5 are the first class constraints called *Hamiltonian constraint* and *momentum constraints*, respectively. We then consider the amplitude of the spatial geometry. A convenient gauge choice is the proper-time gauge:

$$\dot{N} = 0, \quad N_i = 0, \tag{1.2}$$

where the dot means the time derivative. Denoting the initial and final spatial geometries by Σ_1 and Σ_2, those become arguments of the amplitude:

$$A[\Sigma_2, \Sigma_1] = \int [\mathscr{D}f][\mathscr{D}T]\langle h(2)|\exp\Big(-i\int dx\; T\mathscr{H}_{\text{eff}}\Big)|h(f(1))\rangle. \tag{1.3}$$

In the following, we explain ingredients in the amplitude (1.3). $h(1)$ ($h(2)$) is the eigenvalue of the spatial metric h_{ij} on the spatial geometry Σ_1 (Σ_2); the $|h(1)\rangle$ ($|h(2)\rangle$) is the eigenstate of h_{ij} with the eigenvalue $h(1)(h(2))$. f is the spatial diffeomorphism acting on the initial geometry, $x \to f(x)$; $[\mathscr{D}f]$ is its diffeomorphism-invariant measure. dx is an abstract description of the spatial measure. T is the proper time defined as

$$T(x) = (t_2 - t_1)N(x), \tag{1.4}$$

where x is the coordinate of the final spatial geometry Σ_2. The measure $[\mathscr{D}T]$ is defined as the infinite product of $dT(x)/T(x)$. Integrating over $T(x)$ generates all possible locations of the final spatial geometry associated with each initial geometry

[1] As for the notation of constraints, see Sect. 1.3.

$h(f(1))$. The effective Hamiltonian density $\mathscr{H}_{\mathrm{eff}}$ includes the ghost Hamiltonian arising from choosing the proper-time gauge in addition to the original one $\mathscr{H}$. We then put the causality into the amplitude in such a way that Σ_2 lies on the future of Σ_1. This can be realized by restricting the range of T such that

$$T(x) > 0. \tag{1.5}$$

We denote the amplitude with only positive proper time as $A_+[\Sigma_2, \Sigma_1]$. We call it *causal amplitude*. We put aside this problem, but we will again come to this point later. According to Dirac's theory of constrained systems, the Hamiltonian constraint and momentum constraints are generators of the gauge transformation, i.e., diffeomorphism (see Sect. 1.3). Acting the momentum constraint Φ_5^i on A_+, one finds

$$\Phi_5^i A_+[\Sigma_2, \Sigma_1](h(1)) = 0, \quad \text{or} \quad \Phi_5^i A_+[\Sigma_2, \Sigma_1](h(2)) = 0. \tag{1.6}$$

From this, one finds that the causal amplitude is invariant under the spatial diffeomorphism. On the other hand, acting the Hamiltonian constraint on the causal amplitude leads

$$\Phi_4 A_+[\Sigma_2, \Sigma_1](h(1)) \neq 0, \quad \text{or} \quad \Phi_4 A_+[\Sigma_2, \Sigma_1](h(2)) \neq 0. \tag{1.7}$$

Thus, the causal amplitude breaks the gauge invariance associated with the surface deformation (see Sect. 1.3). When restoring the range of the proper time such that $-\infty \leq T \leq +\infty$, one obtains

$$\Phi_4 A[\Sigma_2, \Sigma_1](h(1)) = 0, \quad \text{or} \quad \Phi_4 A[\Sigma_2, \Sigma_1](h(2)) = 0. \tag{1.8}$$

Therefore, we conclude that the causality, Σ_2 lies on the future of Σ_1, can not coexist with the gauge invariance. This is not an unnatural consequence because in the case of quantum mechanics for a relativistic scalar particle in an external field, one finds the similar situation [1].

When taking the causality rather than gauge invariance as a fundamental property of the amplitude of quantum gravity, an introduction of the spatial foliation, the spatial hypersurface labeled by time, is very efficient as we have seen above because one can take in the notion of causality in such a way that each leaf of the foliation never touch each other and flows to the future. This situation can be naturally realized as the Lorentzian lattice quantum gravity called *Causal Dynamical Triangulations* (CDT for short) [3]. CDT partially has answered the question raised above, "*Why is the Universe the way it is?*": summing up all the classical geometries by the CDT method, a de-Sitter universe has been obtained as a low-energy realization [4].

Apart from Teitelboim's argument, there exists another logical possibility to introduce the foliation structure as an observable quantity in quantum gravity. In that case, the Lorentz symmetry ought to be broken at the fundamental level in such a way as to recover it approximately at a long distance scale. One way to realize this situation is to introduce a time-like vector field coupled to gravity, like æther. A well-known example is the Einstein-æther theory ([5] is a good review). At low energies, the time-like vector field, called æther, is effectively decoupled; this is only possible because gravity is so week. In the Einstein-æther theory, the time-like vector field is an external degree of freedom from the point of view of gravity. The n-DBI gravity [6, 7] is in the same spirit. The time-like vector field in n-DBI gravity, n, serves as a *clock* specifying the direction of time. One witching feature of n-DBI gravity is to drive inflation without introducing any scalar field agent.

In this thesis, we explore the physics of space-time with the space-time foliation based on two theories: CDT and n-DBI gravity. It seems that there is no pathological behavior in both theories within our analysis. Hope that readers will enjoy the author's journey to quantum gravity during his Ph.D. life through the thesis.

1.2 Quantum Gravity Without Coordinates

1.2.1 General Relativity Without Coordinates

Field theory carries, in general, infinite number of degrees of freedom and therefore, divergence is its built-in nature. This implies necessity of introducing the cut-off in momentum space Λ, or equivalently the lattice spacing $\varepsilon = \Lambda^{-1}$ as the smallest probing length scale. Imposing the cut-off or lattice spacing can be seen as a coarse graining of degrees of freedom. If the theory is renormalizable, after removing the regulator, i.e., $\Lambda \to \infty$ or $\varepsilon \to 0$, one can obtain renormalized finite physical quantities. In lattice theories, translating the scale defined by the lattice spacing ε to physical quantities, say physical mass, one needs to tune coupling constants to some value where correlation lengths diverge. The correlation length is the smallest length in the unit of lattice spacing which does not change qualitative nature of the system. For instance, if one defines two operators located on lattice cites, $n\varepsilon$ and $m\varepsilon$, as $\phi(n\varepsilon)$ and $\phi(m\varepsilon)$, and considers the large separation, $|n-m| \gg 1$, then the correlation function behaves as

$$\langle \phi(n\varepsilon)\phi(m\varepsilon)\rangle \sim e^{-|n-m|/\xi(g)}, \tag{1.9}$$

where $\xi(g)$ is the correlation length as a function of some coupling constant g. When approaching the critical value of g, say g_c, the correlation length diverges as

$$\xi(g) \propto \frac{1}{|g-g_c|^{\nu}}. \tag{1.10}$$

Simultaneously tuning the lattice spacing in such a way that

$$\varepsilon(g) \propto |g-g_c|^{\nu}, \tag{1.11}$$

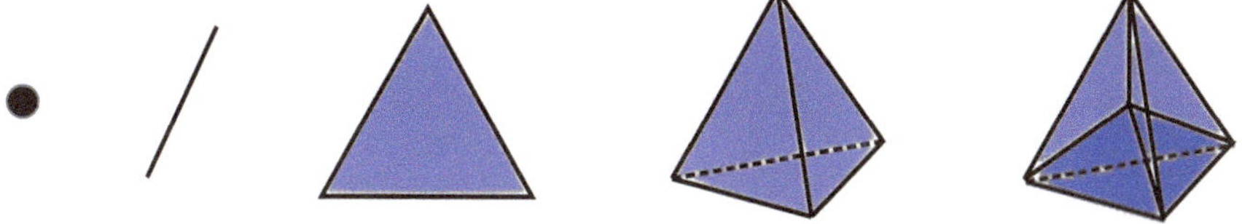

Fig. 1.1 Simplices in several dimensions. Starting from the *left*, these are the 0-simplex (*vertex*), 1-simplex (*edge*), 2-simplex (*triangle*), 3-simplex (*tetrahedra*) and 4-simplex

one can introduce the physical mass, $m_p = 1/\xi(g)\varepsilon(g)$, and the physical length, $|x_n - x_m| = \varepsilon(g)|n - m|$, as fixed quantities when $g \to g_c$. This procedure is called *continuum limit.*[2] In the Wilsonian renormalization group, the point (or more generally co-dimension surface) characterized by the critical coupling constant(s) is the fixed point (surface) under the renormalization group flow. Therefore, in the lattice theory, if one can find the fixed point (surface) of the renormalization group, physical quantities can be extracted at the long-distance scale where the lattice spacing is small enough to be neglected.

General relativity is a field theory, and so its lattice formulation can be anticipated. The lattice formulation of general relativity ought to discribe the dynamics of the space-time lattice. This has been done by Regge in 1961 [10]. He imposed the lattice structure on a curved space-time manifold using simplices and investigated its dynamics. Such regularized manifold is called *simplicial manifold.* Let us explain the basic idea of Regge's formulation of lattice gravity. For simplicity, we work on a Euclidean d-dimensional manifold. In d dimensions, the fundamental building block (lattice structure) is the d-simplex (see Fig. 1.1 for example).

The d-simplex consists of $(d-1)$-, $(d-2)$-, $\cdots$, 1- and 0-simplices. For instance, the 2-simplex (triangle) consists of 1-simplex (edge) and 0-simplex (vertex). All links and faces in the d-simplex are straight and flat, respectively. Discretizing the d-dimensional manifold by d-simplices is called *simplicial decomposition* or *triangulation.* In the d-dimensional triangulation, each building block, d-simplex, holds $(d-1)$-simplices in common with its adjacent d-simplices, and the lattice spacing is the length of 1-simplex (edge). In Regge's formulation, information of the curvature at the scale shorter than the lattice spacing is coarse-grained, since each building block consists of straight edges. And then, how is the curvature described in a simplicial manifold? In general, the curvature is defined by the deviation of a vector under an infinitesimal parallel translation along a closed path. And the translation along the infinitesimal closed path is a notion related to the 2-dimensional plane in any dimensions larger than or equal to 2. Therefore, the curvature in the d-dimensional simplicial manifold is supposed to be measured by the translation of a vector around the co-dimension 2 object, $(d-2)$-simplex. Such a $(d-2)$-simplex is called *hinge*. In order to explain the statement above in detail and read off the curvature structure around the hinge, taking the 4-dimensional case as an example,

[2] See, for instance, [9].

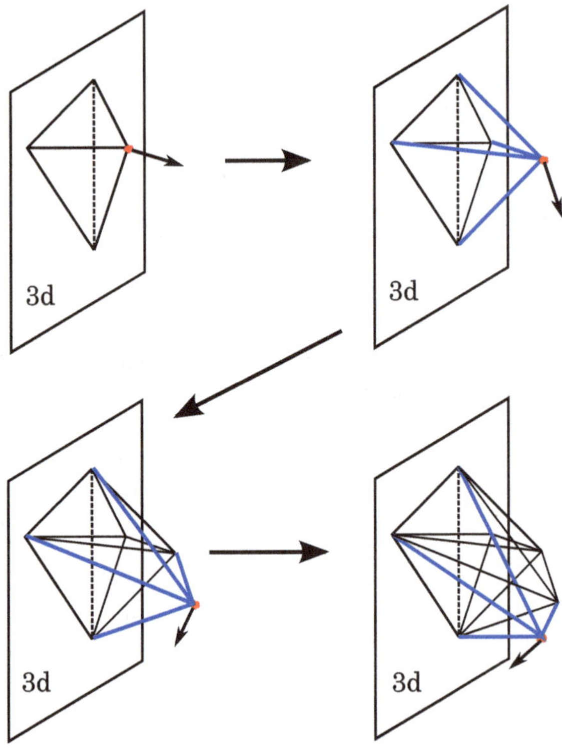

Fig. 1.2 Rotation around the hinge

we reconstruct the simplicial manifold by rotating a vertex around the hinge. See Fig. 1.2. Firstly, we pick up a 2-simplex, which is the hinge in 4 dimensions, and then add a vertex colored red away from the 2-dimensional plane where the hinge is embedded. Thereby, we can connect the vertex colored red with each vertex of the hinge to construct a 3-simplex. This 3-simplex is written in the upper-left of Fig. 1.2. Next, we rotate the red-colored vertex certain degrees around the hinge in a direction toward the extra dimension proportional to the 3-dimensional hyperplane where the 3-simplex is embedded. This situation is described in the upper-right of Fig. 1.2. Through this step, it turns out that we have obtained the 4-simplex. Therefore, in fact the 4-simplex is constructed by a rotation of a vertex around the hinge. Rotating the vertex some degrees repeatedly, one can reconstruct the 4-dimensional simplicial manifold. In addition, depending on how to choose the rotational plane, the curvature seems to be changed and expressed by a conical singularity at the hinge. Generally, in d dimensions, one observes the conical singularity at the location of the hinge by rotating a vertex around the hinge (see Fig. 1.3). The existence of conical singularity implies the deficit angle (see Fig. 1.3). This is a special feature of the simplicial manifold.

Next, let us step into more rigorous discussion and describe geometrical quantities on the lattice. Especially, we will derive the lattice-analogue of the Riemann tensor

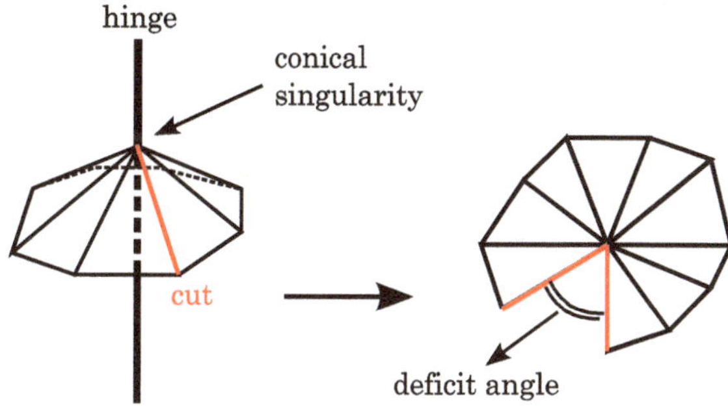

Fig. 1.3 Conical singularity and deficit angle

without coordinate. Again, look at Fig. 1.3. From now, we depict the d-simplex as σ^d. As can be seen from Fig. 1.3, the deficit angle $\delta_{\sigma^{d-2}}$ around the hinge σ^{d-2} is described as follows:

$$\delta_{\sigma_a^{d-2}} = 2\pi - \sum_b \theta(\sigma_b^d, \sigma_a^{d-2}), \tag{1.12}$$

where a and b label simplices; σ_b^d's include σ_a^{d-2} as the sub-simplex; $\theta(\sigma_b^d, \sigma_a^{d-2})$ is the dihedral angle of σ_b^d associated with the hinge σ_a^{d-2}. From now, we try to rewrite the Riemann tensor R_{ijkl} $(i, j, k, l = 1, 2, \cdots, d)$ based on quantities in the simplicial manifold. Since the region outside hinges is flat, we define the inner product by the flat metric and write all tensorial quantities so as to carry lower indices. Firstly, we define the generator of rotation whose rotational plane is orthogonal to the hinge σ_a^{d-2}:

$$S_{ij} = \frac{1}{(d-2)!V_{\sigma_a^{d-2}}} \varepsilon_{ijk_1\cdots k_{d-2}} l_{k_1} \cdots l_{k_{d-2}}, \tag{1.13}$$

where $V_{\sigma_a^{d-2}}$ is the volume of σ_a^{d-2}; $\varepsilon_{ij,k_1\cdots k_{d-2}}$ is an anti-symmetric tensor defined by $\varepsilon_{12\cdots d} = 1$; $l_{k_1}, \cdots, l_{k_{d-2}}$ are vectors shaping the hinge σ_a^{d-2}. The Riemann tensor can be obtained through a parallel translation along an infinitesimal path surrounding the hinge. To define the vector in a rigorous manner on the 2-dimensional plane, we introduce the local continuum limit. See Fig. 1.4. In the d-dimensional simplicial manifold, we take a minimal surface which surrounds one hinge σ_a^{d-2}. Here we consider inserting the same hinges into the minimal surface, and define the density of hinges as $\rho_{\sigma^{d-2}}$. If one takes the limits, $\rho_{\sigma^{d-2}} \to \infty$ and $\delta_{\sigma^{d-2}} \to 0$, simultaneously under the quantity $\rho_{\sigma^{d-2}}\delta_{\sigma^{d-2}}$ is fixed, then the minimal surface becomes smooth. We call this limit *local continuum limit*. On this continuous surface, one can define the vector. We denote this surface and its area element as Σ and Σ_{ij}, respectively. Additionally, we define a closed path C as the boundary of Σ. Here we define the number of hinges inside Σ as follows:

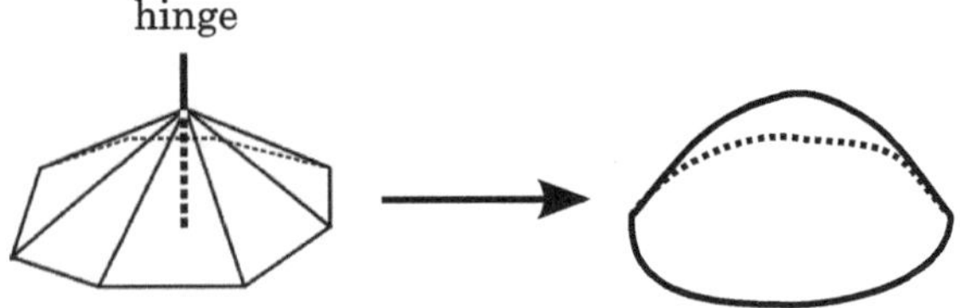

Fig. 1.4 Local continuum limit

$$N_{\sigma^{d-2}} = \frac{1}{2}\rho_{\sigma^{d-2}}\Sigma_{ij}S_{ij}. \tag{1.14}$$

Letting the vector ξ_i go around Σ along the path C, one observes that the vector is rotated $N_{\sigma^{d-2}}\delta_{\sigma^{d-2}}$ degree comparing the configurations of the vector before and after the parallel translation. The infinitesimal variation of ξ_i induced by this parallel translation can be written as follows:

$$\delta\xi_i = N_{\sigma^{d-2}}\delta_{\sigma^{d-2}}S_{ij}\xi_j. \tag{1.15}$$

This quantity can be also written in terms of the Riemann tensor:

$$\delta\xi_i = \frac{1}{2}\Sigma_{lm}R_{lmij}\xi_j. \tag{1.16}$$

Comparing (1.15) and (1.16), one finds the Riemann tensor:

$$R_{ijkl} = \rho_{\sigma^{d-2}}\delta_{\sigma^{d-2}}S_{ij}S_{kl}. \tag{1.17}$$

Remember that the quantity $\rho_{\sigma^{d-2}}\delta_{\sigma^{d-2}}$ has been fixed under the local continuum limit. Therefore, one can rewrite this quantity based on the language in the simplicial manifold, i.e.,

$$R_{ijkl} = \delta^2(\sigma_a^{d-2})\delta_{\sigma_a^{d-2}}S_{ij}S_{kl}, \tag{1.18}$$

where $\delta^2(\sigma_a^{d-2})$ is the 2-dimensional delta-function whose support is located at the hinge σ_a^{d-2}. This expression does not depend on coordinates. Moreover, contracting indices, one finds the Ricci scalar:

$$R = R_{ijij} = 2\delta^2(\sigma_a^{d-2})\delta_{\sigma_a^{d-2}}. \tag{1.19}$$

Finally, we have arrived at the geometric quantities without coordinate: the Riemann tensor and Ricci scalar depend on the location of the hinge. In particular, we again stress that the Ricci scalar is expressed as the conical singularity at the hinge, which is a remarkable property for the simplicial manifold. The lattice formulation of general relativity is quite useful especially for investigating the quantum mechanical nature of gravity. It is widely known that the Einstein-Hilbert action suffers from its non-renormalizability by the power-counting argument. It would be premature that from this argument use of Einstein-Hilbert action yields pathologies. Strong reasons why we stick to the Einstein-Hilbert action is that it ensures the unitarity and "in Riemann's space, R is the sole invariant that contains the derivatives of the $g_{\mu\nu}$ only to the second order (by Weyl [11])." If there exists a non-Gaussian ultraviolet fixed point, then the conventional argument of power counting cannot be applied anymore. The existence of such a non-trivial fixed point has been recently reported by the authors in [12–16], using the exact renormalization group approach. This is called *aymptotic safety scenario of gravity* started by Weinberg [17] and developed by Reuter [18]. An alternative way of searching for such a non-Gaussian ultraviolet fixed point is a non-perturbative quantum gravity on the lattice, especially (causal) dynamical triangulations. Since the lattice regularization allows us to conduct the path-integral non-perturbatively, there is a possibility to find the non-Gaussian fixed point. In the following, we will explain the idea of (causal) dynamical triangulations.

1.2.2 Causal Dynamical Triangulations

Next, let us see the quantum-mechanical formulation of lattice gravities. In general, two possible formulations are known depending on the choice of the dynamical variable. First one is called *Regge calculus* in which the link length becomes dynamical in a fixed triangulation of the manifold, which has been established classically by [10]. Quantum-mechanical formulation of the Regge calculus called *quantum Regge calculus* has been launched by Pozano and Regge [19] and later by Turaev and Viro [20]. This line of study is ongoing as *spin foam model* or *loop quantum gravity*. There is an alternative formulation which is called *dynamical triangulations* (DT) firstly introduced in [21–26]. In DT, the triangulation is chosen as the dynamical variable when link lengths of simplces are fixed as the same lattice spacing ε. This program has been developed into so-called *causal dynamical triangulation* (CDT) started by Ambjørn and Loll [3]. In the following, we will explain the idea of DT and CDT.[3]

To begin, we construct the gravitational action of DT. Remember the Einstein-Hilbert action with Euclidean signature on the manifold $\mathscr{M}$:

$$S = -\frac{1}{16\pi G_d} \int_{\mathscr{M}} d^d x \sqrt{g}\,(R - 2\Lambda)\,, \tag{1.20}$$

where G_d and Λ are the Newton constant and cosmological constant, respectively. Plugging the Ricci scalar (1.19) into the action (1.20), one arrives at the following lattice action with the lattice spacing ε:

$$S_{\mathrm{DT}} = -\frac{1}{16\pi G_d \varepsilon^2} \sum_{\sigma_a^{d-2} \in T} \left(2\delta_{\sigma_a^{d-2}} V_{\sigma_a^{d-2}} - 2\Lambda V(\sigma_a^{d-2}) \right), \tag{1.21}$$

[3] There is a nice review of CDT [9]; in this book, we follow the notation appeared there.

where T is a triangulation of $\mathscr{M}$; $V_{\sigma_a^{d-2}}$ is the volume of σ_a^{d-2}; $V(\sigma_a^{d-2})$ has been defined by the volume of d-simplex V_{σ^d} as

$$V_d(\sigma_a^{d-2}) = \frac{2}{d(d+1)} \sum_{\sigma^d \ni \sigma_a^{d-2}} V_{\sigma^d}. \tag{1.22}$$

In (1.22), the factor $2/d(d+1)$ appears because each d-simplex has $d(d+1)/2$ hinges, so that when summing over $(d-2)$-simplices in (1.21) it turns out one counts V_d redundantly, and the redundant number is $d(d+1)/2$. The action (1.21) is called *Regge action*. In DT, the path-integral of the metric g weighted by the Einstein-Hilbert action S,

$$Z = \int \frac{1}{V(\mathrm{Diff})}[\mathscr{D}g]e^{-S[g]}, \tag{1.23}$$

is replaced by the sum over triangulations T weighted by the Regge action S_{DT}:

$$Z_\varepsilon = \sum_T \frac{1}{C_T} e^{-S_{\mathrm{DT}}[T]}, \tag{1.24}$$

where $V(\mathrm{Diff})$ is the gauge volume of the diffeomorphism group; C_T is the order of the automorphism group of T; ε in Z_ε is the lattice spacing of T. In 4 dimensions, DT could not produce any physically interesting phase at least without matter fields.

So far, we argued in the Euclidean setup. From now, we shall introduce CDT as a Lorentzian realization of DT. Firstly, we define the Lorentzian version of Regge action:

$$S_{\mathrm{CDT}} = -\frac{1}{16\pi G_d \varepsilon^2} \sum_{\sigma_a^{d-2} \in T} \left(2\delta_{\sigma_a^{d-2}} V_{\sigma_a^{d-2}} - 2\Lambda V(\sigma_a^{d-2})\right), \tag{1.25}$$

where

$$\delta_{\sigma_a^{d-2}} = \left(2\pi - \sum_b \theta(\sigma_b^d, \sigma_a^{d-2})\right) e^{i\phi(\sigma^{d-2})}. \tag{1.26}$$

We will explain ingredients in (1.25). Firstly, a Lorentzian d-simplex is defined in the following way: one prepares simplices in Euclidean space and then connects two Euclidean simplices by time-like vectors so that it forms a d-simplex. See Fig. 1.5. Red and black lines indicate time-like and space-like edges. Lengths of time-like and space-like edges, l_L and l_E, are defined in such a way that $l_L^2 = -\alpha\varepsilon^2$, $l_E^2 = \varepsilon^2$. Here α is a positive parameter. All ingredients in (1.25) coincides with those in Euclidean case except for two following things. Firstly, a volume of each simplex is computed using time-like vectors for time-like edges and space-like vectors for space-like edges. Secondary, the deficit angle (1.26) has a phase $\phi(\sigma^{d-2})$. ϕ has been defined so that $\phi = 0$ for space-like hinges, and $\phi = -\pi/2$ for time-like hinges.

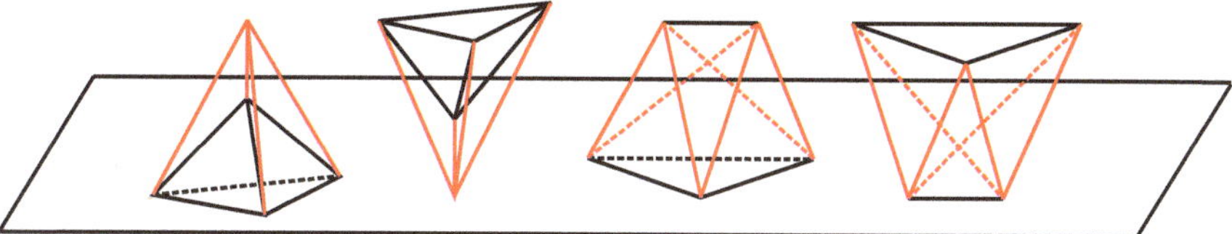

Fig. 1.5 4-simplices with Lorentzian signature

This phase factor ensures the reality of deficit angle. A partition function of CDT is then defined as

$$Z_{\varepsilon} = \sum_{T} \frac{1}{C_T} e^{i S_{\mathrm{CDT}}[T;\alpha]}. \tag{1.27}$$

Because of a reality condition for volume of simplices, the parameter α is confined to some range [27]. We summarize three important requirements in CDT. First, one discretizes the space-time using simplices with Lorentzian signature. Second, a proper-time slicing is imposed on the regularized space-time in such a way that time does not go backwards. Third, one prohibits spatial topology change. The first requirement ensures that the space-time has Lorentzian signature. Each geometry in the sum should be a classical discretized geometry as each trajectory in the path-integral of a relativistic point particle is a classical path [1]. Since it is natural that each classical discretized geometry have the causality built-in according to Teitelboim's argument, the second requirement has been imposed. The third requirement means that in CDT baby universe contributions are integrated out. Without this requirement, one observes infinite creation of baby universes at instant time, which causes problems at low energies. Although at first sight this third requirement seems to be strong, at least in $1+1$ dimensions, it has been shown that one can relax it to allow for a CDT with the creation of baby universes [28]. Thus, the third requirement could be removed for $3+1$ dimensions as well. One attractive feature of CDT is that one can implement the discrete-analogue of path-integral non-perturbatively. This makes it possible to find a non-Gaussian fixed point where the conventional power counting argument cannot be applied. In $(3+1)$-dimensional CDT, at low energies a de Sitter Universe has been obtained via computer simulations [4].

For running computer simulations, one needs a well-defined Boltzmann weight. In CDT, one can map a Lorentzian simplicial manifold to a Euclidean one changing α to $-\alpha$. This procedure looks like a conventional Wick rotation, but this map is a strict bijection between Lorentzian and Euclidean simplicial geometries. Through this map, the partition function can be altered into a Euclidean one:

$$Z_{\varepsilon} = \sum_{T} \frac{1}{C_T} e^{-S^{(e)}_{\mathrm{CDT}}[T,-\alpha]}, \tag{1.28}$$

where $S^{(e)}_{\mathrm{CDT}}[T; -\alpha]$ is a Euclidean action with negative value of α.[4] This partition function is different from the partition function of DT, i.e., each simplicial geometry has the causality built-in. Euclidean quantum gravity suffers from the action unbounded from below. The CDT Euclidean action $S^{(e)}_{\mathrm{CDT}}[T; -\alpha]$ is also unbounded from below, but it can be cured by *entropic* reasons. To see what it is and how it works, we replace the sum over triangulations by the sum over simplicies in the partition function:

$$\begin{aligned} Z_\varepsilon &= \sum_N e^{-S^{(e)}_{\mathrm{CDT}}[N,-\alpha]} \sum_{T(N)} \frac{1}{C_{T(N)}} \\ &= \sum_N e^{-S^{(e)}_{\mathrm{CDT}}[N,-\alpha]+\log w_N}, \end{aligned} \tag{1.29}$$

where N means the number of every kind of simplex in a simplicial manifold, and w_N is the number of triangulations under a fixed N. The logarithm of w_N is an entropy for a fixed N. This entropy factor is designed to suppress rare configurations in the sum. In CDT, a large negative value of the action is really suppressed by entropic reasons in the continuum limit, $\varepsilon \to 0$ [29].

1.2.3 Exactly Solvable Models

We explain the $(1 + 1)$-dimensional model of CDT. Since the $(1 + 1)$-dimensional CDT can be solved analytically, one can pointedly understand basics and unsolved problems of CDT in some depth through this toy model. As explained in Sect. 1.2.2, in CDT there is an exact map between individual Euclidean and Lorentzian simplicial geometries. Therefore, from now on we work on Euclidean geometries with the causality built-in. First of all, 2-dimensional gravity has no degrees of freedom. This fact implies that 2-dimensional gravity is *topological*. Integrals of curvature in both continuous and discrete geometries become a constant called *Euler number*:

$$\int_{\mathcal{M}} d^2x \sqrt{g} R = 4\pi\chi(\mathcal{M}) = 4\pi\chi(T) = \sum_{\sigma^0_a \in T} 2\delta_{\sigma^0_a} V_{\sigma^0_a}, \tag{1.30}$$

where $\chi(\mathcal{M})$ and $\chi(T)$ are the Euler number of a manifold $\mathcal{M}$ and triangulation T. This is called *Gauss-Bonnet theorem*. Introducing matter fields, one can see non-trivial dynamics. Polyakov succeeded in formulating a non-trivial 2-dimensional gravity integrating out scalars minimally coupled to gravity [32]. This theory is called *Liouvulle field theory*.[5] In the Liouville field theory, the number of scalers

[4] In 2 dimensions the difference of α can be absorbed into the redefinition of the cosmological constant, so that one can set $\alpha = 1$ without loss of generality.

[5] For a good review of the Liouville field theory, see [33].

c, *central charge* of matters, is a parameter built-in. Setting the parameter as zero, observables in the Liouville field theory coincide with those of DT in the continuum limit. Considering matter-coupled DTs, it holds up to $c = 1$. DT coupled to matters can be constructed using a powerful tool called *matrix model*.

In the following, we show that DT and CDT can be unified in 2 dimensions firstly pointed out in [34]. This is only possible because configurations of CDT are included in those of DT. First, we simply investigate DT by combinatorics and then extract information of CDT. We start with the following boundary-to-vacuum amplitude with sphere topology:

$$A(l;\lambda) = \sum_{T_l} \frac{1}{C_{T_l}} e^{-S_{\mathrm{DT}}[T_l]}, \tag{1.31}$$

where

$$S_{\mathrm{DT}} = \lambda n(T_l). \tag{1.32}$$

In the above, we have eliminated ε-dependence introducing a dimensionless bulk cosmological constant λ; we have denoted the number of triangles in the triangulation with fixed boundary edges T_l and the number of boundary edges by $n(T_l)$ and l, respectively. Because of the Gauss-Bonnet theorem, the curvature term is trivial. Therefore, we ignored it. Although it is possible to sum over different kinds of topologies in the partition function, it might cover what we stress on here. We thus focus only on sphere topology. For computational reasons, we introduce a generating function of the boundary:

$$w(\lambda,\mu) = \sum_{l=1}^{\infty} A(l;\lambda) e^{-\mu l}, \tag{1.33}$$

where μ is a boundary cosmological constant. Changing the sum over triangulations to the sum over the number of triangles, one finds

$$w(g,z) = \sum_{n=0}^{\infty}\sum_{l=0}^{\infty} w_{n,l}\, g^n z^{-l-1}, \tag{1.34}$$

where $g = e^{-\lambda}$, $z = e^{\mu}$ and we have redefined l so that it starts from 0. In the above, $w_{n,l}$ is the number of triangulations with fixed n and l. Information distinguishing DT and CDT should be included in $w_{n,l}$. Remember that important things to construct CDT are the proper-time slicing and the suppression of baby universes. To extract CDT configurations from the full DT configurations, we introduce a 2-gon called *double link* consists of 2 edges: we discretize geometries using triangles and double links as building blocks. Such an extended triangulation is called *unrestricted triangulations*. Since ordinary triangulations and unrestricted triangulations yield same physics in the continuum limit, *universality*, one can use unrestricted triangulations without loss of generality. Marking a point on one of boundary edges, one can find

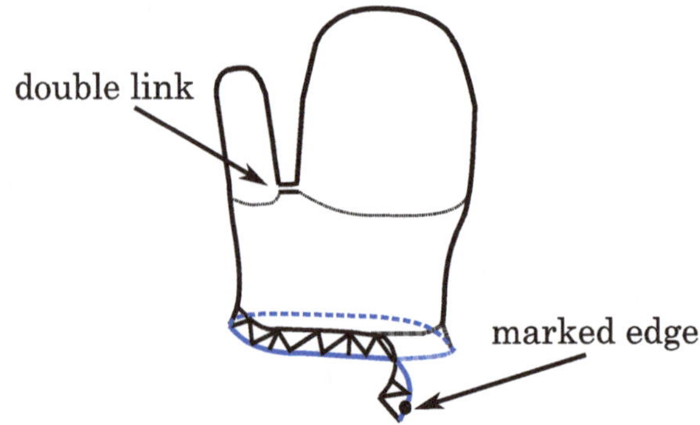

Fig. 1.6 Peeling method

a recursion relation:

$$\left[w(g,z)-\frac{w_0(g)}{z}\right]=gz\left[w(g,z)-\frac{w_0(g)}{z}-\frac{w_1(g)}{z^2}\right]+\frac{1}{z^2}\left[zw(g,z)^2\right], \quad (1.35)$$

where

$$w_m(g)=\sum_{n=0}^{\infty} w_{n,m}g^n. \quad (1.36)$$

This equation means that when marking a point on one of boundary edges there are two possibilities: a marked edge belongs to a triangle or double link. The left-hand side of (1.35) stands for all possible configurations. The term $\frac{w_0(g)}{z}$ is subtracted because if there is no edge, this recursion relation does not hold. The first term in the right-hand side of (1.35) means the case that a marked edge belongs to a triangle. The terms, $\frac{w_0(g)}{z}$ and $\frac{w_1(g)}{z^2}$, are subtracted because if there is no edge and triangle, this recursion does not hold. The second term of the right-hand side of (1.35) means the case that a marked edge belongs to a double link. If one peels triangles or double links attached to the boundary like peeling an apple, one can obtain a new boundary and construct the same kind of recursion relation. One can consider new boundaries arising from successive peeling procedure as proper-time slicing in CDT without loss of generality. Doing so, spatial topology change only happens if one encounters a double link in the peeling. Therefore, imposing a weight on double links, one can control a creation of baby universes (see Fig. 1.6). We denote such a wight by g_s and call it *string coupling constant*. Adding this weight to the recursion relation (1.35), one finds (see Fig. 1.7):

$$\left[w(g,z)-\frac{w_0(g)}{z}\right]=gz\left[w(g,z)-\frac{w_0(g)}{z}-\frac{w_1(g)}{z^2}\right]+\frac{g_s}{z^2}\left[zw(g,z)^2\right]. \quad (1.37)$$

Setting $w_0(g)=1$ as normalization, one can rewrite (1.37) as

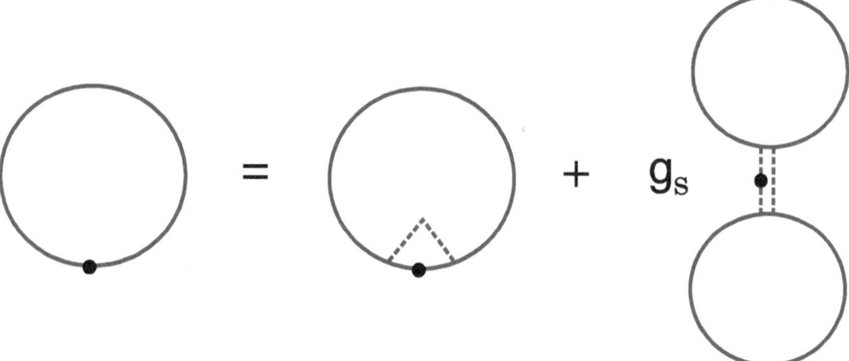

Fig. 1.7 Loop equation. Each term in (1.37) corresponds to each graph in this figure

$$g_s w(g,z)^2 = V'(z)w(g,z) - Q(z), \tag{1.38}$$

where

$$V(z) = \frac{1}{2}z^2 - \frac{g}{3}z^3, \quad Q(z) = 1 - g(w_1(g) + z). \tag{1.39}$$

(1.38) is called *loop equation*. More generally, one can discretize geometries using any polygons: 1-gon, 2-gon, ..., n-gon. Corresponding loop equation is equivalent to (1.38) except that

$$V(z) = \frac{1}{2}z^2 - g\sum_{m=1}^{n}\frac{t_m}{m}z^m, \quad Q(z) = 1 - g\sum_{j=2}^{n}t_j\sum_{m=0}^{j-2}z^m w_{j-2-m}(g), \tag{1.40}$$

where $t_m g$ is the weight of m-gons. For technical reasons, we extend a real z to a complex variable. The solution of the loop equation is then obtained

$$w(g,z) = \frac{V'(z) - \sqrt{V'(z)^2 - 4g_s Q(z)}}{2g_s}, \tag{1.41}$$

where the minus sign in front of the square root has been chosen such that $w(g,z)$ asymptotes to $1/z$ for large $|z|$ (see (1.34)). An important point here is that the following quantity is a polynomial of degree $n-1$:

$$\sigma(z) = \sqrt{V'(z)^2 - 4g_s Q(z)}. \tag{1.42}$$

We assume that branch cuts are located on the real axis and $w(z)$ is an analytic function in the complex z-plane except for vicinity of the cuts. One can then choose the branch-cut structure of $\sigma(z)$:

$$\sigma(z) = M(z)\prod_{i=1}^{n-k}\sqrt{(z-c_{i+})(z-c_{i-})}, \quad (k = 1, 2, \cdots, n-1), \tag{1.43}$$

where $M(z)$ is a polynomial of degree $k-1$, and c_{i+} and c_{i-} are end points of the cuts such that $c_{i+} > c_{i-}$. The asymptotic behavior of $w(z)$ in $|z| \gg c_{i+} - c_{i-}$ is powerful enough to determine all coefficients in $M(z)$, c_{i+} and c_{i-} as functions of g_s and $t_m g$'s. To see critical behaviors of $w(g, z)$, one needs to focus on zeros of $M(z)$: one can see critical behaviors when some roots of $M(z)$ approach to c_{i+} or c_{i-}. At such critical points, differentiating c_{i+} or c_{i-} w.r.t. g yields a singularity. In the following discussion, we focus on a single cut for simplicity. To explain the idea of critical phenomena in both DT and CDT and how the continuum limit can be taken, we introduce efficient variables called *moments* [30]:

$$M_k = \oint_C \frac{d\omega}{2\pi i}\frac{V'(\omega)}{(\omega-c_+)^{k+1/2}(\omega-c_-)^{1/2}}, \tag{1.44}$$

$$J_k = \oint_C \frac{d\omega}{2\pi i}\frac{V'(\omega)}{(\omega-c_+)^{1/2}(\omega-c_-)^{k+1/2}}, \tag{1.45}$$

where $k \geq 1$ and C is a contour enclosing the branch cut (see Fig. 1.8). Remember that $V(z)$ is a polynomial of degree n. For $k \geq n$, one can move the integration contour C to infinity and obtains $M_k = J_k = 0$ for $k \geq n$. Since $w(g, z)$ asymptotes to $1/z$ for $|z| \gg c_+ - c_-$, one finds

$$M(z) = \oint_{C_{\text{inf}}} \frac{d\omega}{2\pi i}\frac{M(\omega)}{\omega - z} = \oint_{C_{\text{inf}}} \frac{d\omega}{2\pi i}\frac{V'(\omega)}{\omega - z}\frac{1}{\sqrt{(\omega-c_+)(\omega-c_-)}}, \tag{1.46}$$

where C_{inf} is a contour enclosing the branch cut and the point z (see Fig. 1.8). From (1.46), one finds that $M(z)$ can be expanded in terms of moments:

$$M(z) = \sum_{k=1}^{n-1} M_k(z-c_+)^{k-1} = \sum_{k=1}^{n-1} J_k(z-c_-)^{k-1}. \tag{1.47}$$

Expanding $\sigma(z)$ by M_k, one finds

$$\sigma(z) = \sum_{k=1}^{n-1} M_k(z-c_+)^{k-1}\sqrt{(z-c_+)(z-c_-)}. \tag{1.48}$$

Setting,

$$M_1 = 0, \tag{1.49}$$

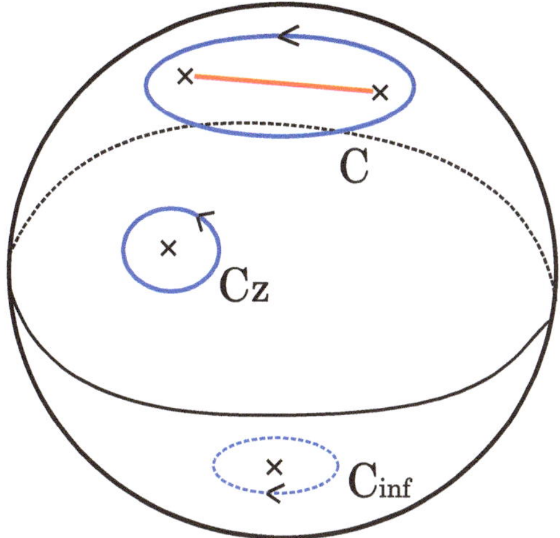

Fig. 1.8 Contours enclosing the cut, z and infinity on the complex ω-plane

the root of $M(z)$ approaches to the end point of the branch cut, c_+. Let us summarize what we have seen so far. Firstly, we have imposed the following requirements (for a single branch cut):

1. $w(g, z)$ asymptotes to $1/z$ in $|z| \gg c_+ - c_-$.
2. $w(g, z)$ is analytic except for vicinity of the branch cut.
3. The branch cut is located on the real axis in the complex z-plane.

We then have found the solution:

$$w(g, z) = \frac{1}{2g_s}\left(V'(z) - \sum_{k=1}^{n-1} M_k (z - c_+)^{k-1}\sqrt{(z - c_+)(z - c_-)}\right), \tag{1.50}$$

where all coefficients in principle can be obtained via the requirement 1 as functions of coupling constants of polygons and g_s. Setting $M_1 = 0$, one can find the critical point where $w(g, z)$ becomes singular after differentiating w.r.t. g with suitable times. Tuning coupling constants but g_s to their critical values and moving z to the end point of the branch cut ($w(g, z)$ is non-analytic for vicinity of the branch cut), one can obtain the disk function of the Liouville field theory. This is a well-established discrete-vs.-continuum structure of DT and the Liouville field theory. One can, in fact, obtain another scaling limit corresponding to the continuum limit of CDT by tuning g_s as well [31]. We will close this section by giving examples of two kinds of continuum limits for DT and CDT.

Continuum Limit of DT

We pick up the following simple $V(z)$:

$$V(z) = \frac{1}{2}z^2 - \frac{g}{4}z^4. \tag{1.51}$$

From now on we call $V(z)$ *potential*. We choose $g_s = 1$ without loss of generality. Since the potential is symmetric, $V(z) = V(-z)$, one obtains

$$w(g, z) = \frac{1}{2}\left(V'(z) - \sum_{k=1}^{3} M_k(z-c)^{k-1}\sqrt{z^2-c^2}\right). \tag{1.52}$$

Considering the asymptotic expansion of $w(g, z)$ in $|z| \gg 2c$, one finds the following equations:

$$M_1 = -\frac{3}{2}c^2 g + 1, \quad M_2 = -2cg, \quad M_3 = -g, \tag{1.53}$$

$$3gc^4 - 4c^2 + 16 = 0. \tag{1.54}$$

Plugging values of moments (1.53) into (1.52), one obtains

$$w(g, z) = \frac{1}{2}\left[z - gz^3 + \left(\frac{1}{2}c^2 g + gz^2 - 1\right)\sqrt{z^2-c^2}\right]. \tag{1.55}$$

Solving (1.54) with respect to c^2, one gets

$$c^2 = \frac{2 - 2\sqrt{1-12g}}{3g}, \tag{1.56}$$

where we have chosen the double sign such that c^2 is analytic for $g = 0$. The critical coupling g_c can be obtained plugging (1.56) into the equation, $M_1 = 0$:

$$g_c = \frac{1}{12}. \tag{1.57}$$

In addition, the critical value of end point of the cut becomes $c_c = c(g_c) = 2\sqrt{2}$. To take the continuum limit, one ought to tune g and z to their critical values as follows:

$$g = g_c e^{-\Lambda\varepsilon} \to \frac{1}{12}(1 - \Lambda\varepsilon^2), \quad z = c_c e^{Z\varepsilon} \to 2\sqrt{2}\left(1 + \frac{\varepsilon}{12}Z\right), \tag{1.58}$$

where ε is the lattice spacing, Λ and Z are the bulk and boundary renormalized cosmological constants. Plugging the fine-tuned values into $w(g, z)$, one obtains

$$w(g, z) = \text{non-scaling terms} + \frac{2}{3}\varepsilon^{3/2} W_{\mathrm{DT}}(Z) + \mathcal{O}(\varepsilon^{5/2}), \quad (1.59)$$

where

$$W_{\mathrm{DT}}(Z) = (Z - \sqrt{\Lambda})\sqrt{Z + \sqrt{\Lambda}}. \quad (1.60)$$

W_{DT} is the continuum limit of the generating function of DT. This is noting but the disk function in the Liouville field theory. An important point is that the potential term does not scale at all in DT.

Continuum limit of CDT

We start with the following potential:

$$V(z) = \frac{1}{2}z^2 - gz - \frac{1}{3}gz^3. \quad (1.61)$$

We have chosen 1-gons, double links and triangles as building blocks of the regularized geometry. An inclusion of the 1-gons as building blocks allows us to take the continuum limit easier and nothing more. One then finds the following generating function:

$$w(g, z) = \frac{1}{2g_s}\left(V'(z) - \sum_{k=1}^{2} M_k (z - c_+)^{k-1} \sqrt{(z - c_+)(z - c_-)} \right). \quad (1.62)$$

From the asymptotic behavior of $w(g, z)$ for $|z| \gg c_+ - c_-$, one obtains a set of equations:

$$M_1 = 1 - \frac{g}{2}(3c_+ + c_-), \quad M_2 = -g, \quad (1.63)$$

$$g_s = \frac{1}{16}\left[M_1 (c_+ - c_-)^2 + \frac{1}{2} M_2 (c_+ - c_-)^3 \right], \quad (1.64)$$

$$g = \frac{1}{2} M_1 (c_+ + c_-) + \frac{1}{8} M_2 (c_-^2 - 6c_+c_- - 3c_+^2). \quad (1.65)$$

Remember that g_s weights creation of baby universes (double links). If one can take the critical point such that the creation of baby universes is suppressed, the arising continuum theory is expected to be CDT. Let's see how it works. We set the following condition to obtain the critical values:

$$M_1 = g_s = 0. \tag{1.66}$$

From this condition, one finds the critical values:

$$g_* = g_c(g_s = 0) = \frac{1}{2}, \quad c_* = c_{c\pm}(g_s = 0) = 1. \tag{1.67}$$

At the critical point, one finds that $c_+ = c_-$. See (1.64). If two end points of the cut are approaching each other (the cut-length shrinks to zero), $g_s \to 0$. One then tunes g_s such that

$$g_s = \varepsilon^3 G_s, \tag{1.68}$$

where G_s is the renormalized string coupling constant. Solving (1.65), we can derive sub-leading terms of g and $c_\pm$ around the critical values in the perturbation of g_s:

$$g_c(g_s) = g_* - \frac{3}{4} G_s^{2/3} \varepsilon^2 + \cdots, \quad c_c(g_s) = c_* + G_s^{1/3} \varepsilon + \cdots. \tag{1.69}$$

We then renormalize g and z as follows:

$$g = g_c(g_s) - \varepsilon^2 \Lambda, \quad z = c_c(g_s) + \varepsilon Z, \tag{1.70}$$

where Λ and Z are conventional renormalized bulk and boundary cosmological constants. Plugging these fine-tuned quantities into $w(g, z)$, the continuum generating function can be obtained [31]:

$$w(g, z) = \varepsilon^{-1} W_{\text{GCDT}}(Z_{\text{cdt}}) + \mathcal{O}(\varepsilon^0), \tag{1.71}$$

where

$$W_{\text{GCDT}}(Z_{\text{cdt}}) = \frac{\Lambda_{\text{cdt}} - \frac{1}{2} Z_{\text{cdt}}^2 + (Z_{\text{cdt}} - H)\sqrt{(Z_{\text{cdt}} + H)^2 - \frac{4G_s}{H}}}{2G_s}, \tag{1.72}$$

and

$$\Lambda_{\text{cdt}} = \Lambda + \frac{3}{2} G_s^{2/3}, \quad Z_{\text{cdt}} = Z + G_s^{1/3}, \quad 2\Lambda_{\text{cdt}} H - H^3 = 2G_s. \tag{1.73}$$

Λ_{cdt} and Z_{cdt} are renormalized bulk and boundary cosmological constants in CDT. The subscript, GCDT, in W_{GCDT} stands for *generalized CDT*. This is because W_{GCDT} is the continuum generating function allowing *mild* spatial topology change weighted by G_s, but it lives in a universality class different from the Liouville field theory. We call such a theory *generalized CDT*. To obtain the CDT result, one needs to take the limit, $G_s \to 0$. Under this limit, one finds the CDT generating function:

$$W_{\mathrm{GCDT}}(Z_{\mathrm{cdt}}) \to W_{\mathrm{CDT}}(Z_{\mathrm{cdt}}) = \frac{1}{Z_{\mathrm{cdt}} + \sqrt{2\Lambda_{\mathrm{cdt}}}}. \quad (1.74)$$

This result coincides with the CDT generating function derived in terms of the so-called *transfer matrix approach* [3].

1.3 Geometrodynamics

1.3.1 ADM Formalism

Diffeomorphism invariance is a guiding principle of Einstein's general relativity. This means that when transforming coordinates at individual space-time points, its physics is unchanged. Therefore, diffeomorphism is a local transformation, i.e., gauge redundancy. Because of the gauge invariance, true degrees of freedom of the graviton reduce to 2 in $3 + 1$ dimensions. In the following, to squeeze out gauge degrees of freedom, we introduce the canonical formulation of general relativity.

In the canonical formalism, one needs to specify the direction of time. One intuitive and instructive approach started by Arnowitt, Deser and Misner is called *ADM decomposition* [35, 36]. This ADM decomposition of space-time allows us to pick up one specific time direction without breaking diffeomorphism invariance. Namely, how to choose time is nothing but the gauge redundancy. We explain the idea and explore the dynamics. We start with a space-time manifold $\mathcal{M}$ equipped with a coordinate (t, x^i) where $i = 1, 2, 3$, and the geodesic distance is measured by the metric $g_{\mu\nu}$. We then introduce a time-like normal vector field n^μ. This can be realized by the following condition:

$$g_{\mu\nu}n^\mu n^\nu = -1. \quad (1.75)$$

We decompose the $\mathcal{M}$ into the direct product space $\Sigma \times R$ in such a way that the Σ is orthogonal to n^μ. An orthogonality can be defined because we have introduced the metric $g_{\mu\nu}$:

$$n^\mu g_{\mu i} = 0. \quad (1.76)$$

Here Σ is the spatial hypersurface characterized by t. We then define the induced metric on the Σ:

$$h_{\mu\nu} = g_{\mu\nu} + n_\mu n_\nu. \quad (1.77)$$

If one picks up a space-time point P, then at the vicinity of P, one can define basis vectors (∂_t, ∂_i) so as to satisfy the following equation:

$$\partial_t = N n^\mu \partial_\mu + N^i \partial_i, \quad (1.78)$$

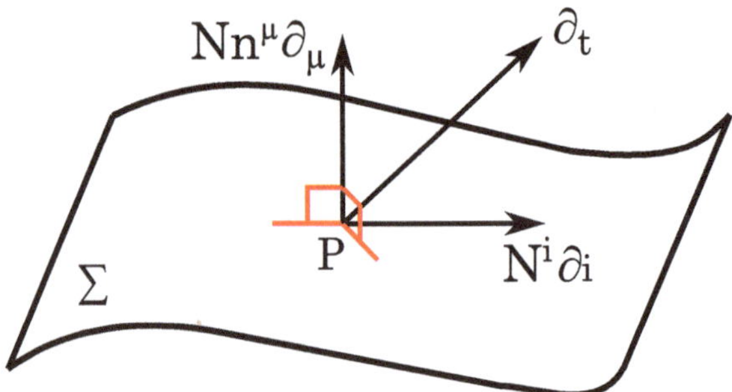

Fig. 1.9 ADM decomposition

where N and N^i are called *lapse function* and *shift vector*, respectively. Using this notation of basis vectors, one can write the metric as follows (see Fig. 1.9):

$$g_{\mu\nu} = \begin{bmatrix} g(\partial_t, \partial_t) & g(\partial_t, \partial_j) \\ g(\partial_i, \partial_t) & g(\partial_i, \partial_j) \end{bmatrix} = \begin{bmatrix} -N^2 + h_{mn}N^m N^n & h_{mj}N^m \\ h_{im}N^m & h_{ij} \end{bmatrix}. \tag{1.79}$$

We introduce a device to measure how the Σ is embedded in the $\mathscr{M}$. It is called *extrinsic curvature* defined as

$$K_{\mu\nu} = \frac{1}{2}\pounds_n h_{\mu\nu}, \tag{1.80}$$

where $\pounds_n$ is the Lie derivative for the vector n^μ, and in the ADM parametrization, it can be written as

$$\pounds_n = \frac{1}{N}(\pounds_t - \pounds_N), \tag{1.81}$$

where $\pounds_t$ and $\pounds_N$ are the Lie derivatives for time and the shift vector. Therefore, the extrinsic curvature can be written like

$$K_{ij} = \frac{1}{2N}(\dot{h}_{ij} - \nabla_i N_j - \nabla_j N_i), \tag{1.82}$$

where the dot means the time derivative and ∇_i is the covariant derivative associated with the h_{ij}. As is clear from the definition (1.80), the extrinsic curvature captures how the spatial metric responds to the change along the vector n^μ orthogonal to the hypersurface. Namely, the shape of the hypersurface (foliation) is determined by the extrinsic curvature. Remember the Einstein-Hilbert action with the cosmological constant term:

$$S = -\frac{1}{16\pi G_4}\int_{\mathscr{M}} d^4x\sqrt{-g}\left({}^{(4)}R - 2\Lambda\right), \tag{1.83}$$

where G_4; ${}^{(4)}R$ and Λ are the Newton constant, the Ricci scalar and the cosmological constant. In the ADM formalism, this action can be written as the following form up to total derivative terms:

$$S_{\text{ADM}} = \int dt L_{\text{ADM}} = -\frac{1}{16\pi G_4} \int\limits_{\Sigma \times R} d^4x N\sqrt{h}(K^{ij}K_{ij} - K^2 + R - 2\Lambda), \quad (1.84)$$

where $K = h^{ij}K_{ij}$; the R is the 3-dimensional Ricci scalar associated with h_{ij}; L_{ADM} is the Lagrangian. We completed our preparations for the canonical formalism. In passing to the Hamiltonian formalism, we set the notation for the canonical conjugate momenta as

$$L_{\text{ADM}}\left((h_{ij}, \dot{h}_{ij}), (N, \dot{N}), (N_i, \dot{N}_i)\right) \to H_{\text{ADM}}\left((h_{ij}, p^{ij}), (N, p_N), (N_i, p^i_{\mathbf{N}})\right). \quad (1.85)$$

However, the time derivatives of N and N_i are absent . Thus we have the primary constraints,

$$\Phi_1 \equiv p_N = 0, \quad \Phi_2^i \equiv p^i_{\mathbf{N}} = 0. \quad (1.86)$$

Denoting the Lagrangian density by $\mathscr{L}_{\text{ADM}}$, the Hamiltonian density is given by

$$\begin{aligned}\mathscr{H}^{(0)}_{\text{ADM}} &\equiv p^{ij}\dot{h}_{ij} - \mathscr{L}_{\text{ADM}} \\ &= \sqrt{h}N_j\left(-\frac{2}{\sqrt{h}}\nabla_i p^{ij}\right) + \frac{\sqrt{h}N}{\kappa}\left[-\frac{\kappa^2}{h}\left(p^{ij}p_{ij} - \frac{1}{2}p^2\right) + R - 2\Lambda\right],\end{aligned} \quad (1.87)$$

where

$$p^{ij} \equiv \frac{\delta L_{\text{ADM}}}{\delta \dot{h}_{ij}} = -\frac{\sqrt{h}}{\kappa}(K^{ij} - h^{ij}K), \quad (1.88)$$

and $\kappa = 16\pi G_4$. The basic non-vanishing Poisson brackets are given by

$$\{p^{ij}(y), h_{kl}(x)\} = \frac{1}{2}(\delta^i_k\delta^j_l + \delta^i_l\delta^j_k)\delta(y - x), \quad (1.89)$$

$$\{p_N(y), N(x)\} = \delta(y - x), \quad (1.90)$$

$$\{p^i_{\mathbf{N}}(y), N_j(x)\} = \delta^i_j\delta(y - x). \quad (1.91)$$

The time flow of the constraints are generated by the extended Hamiltonian density

$$\mathscr{H}^{(1)}_{\text{ADM}} = \mathscr{H}^{(0)}_{\text{ADM}} + \lambda_1\Phi_1 + \lambda_{2i}\Phi_2^i, \quad (1.92)$$

where λ_1 and λ_{2i} are the Lagrange multipliers.

Thus, the primary constraints evolve in time as

$$\begin{aligned}\dot{\Phi}_1(x) &= \int d^3y\{\mathscr{H}^{(0)}_{\mathrm{ADM}}(y), \Phi_1(x)\} + \int d^3y\{\Phi_1(y), \Phi_1(x)\}\lambda_1 \\ &\quad + \int d^3y\{\Phi^i_2(y), \Phi_1(x)\}\lambda_{2i} \\ &= -\frac{\sqrt{h}}{\kappa}\left[-\frac{\kappa^2}{h}\left(p^{ij}p_{ij} - \frac{1}{2}p^2\right) + R - 2\Lambda\right] \equiv \Phi_4(x), \end{aligned} \tag{1.93}$$

$$\begin{aligned}\dot{\Phi}^i_2(x) &= \int d^3y\{\mathscr{H}^{(0)}_{\mathrm{ADM}}(y), \Phi^i_2(x)\} + \int d^3y\{\Phi_1(y), \Phi^i_2(x)\}\lambda_1 \\ &\quad + \int d^3y\{\Phi^j_2(y), \Phi^i_2(x)\}\lambda_{2j} \\ &= 2\nabla_j p^{ij} \equiv \Phi^i_5(x), \end{aligned} \tag{1.94}$$

Therefore, in addition to the primary constraints (1.86), we have the secondary constraints[6]

$$\Phi_4 = \dot{p}_N \approx 0, \ \ \Phi^i_5 = \dot{p}^i_{\mathbf{N}} \approx 0. \tag{1.95}$$

Herein $\approx$ means 'weakly equal' as standard in Dirac's theory. One notices that the Hamiltonian density can be written by constraints:

$$\mathscr{H}^{(1)}_{\mathrm{ADM}} = -(N\Phi_4 + N_i\Phi^i_5) + \lambda_1\Phi_1 + \lambda_{2i}\Phi^i_2. \tag{1.96}$$

It is clear that the Φ_1 and Φ^i_2 commute with all constraints. From the explicit computations, one finds

$$\{\Phi_4(y), \Phi_4(x)\} = \Phi^k_5(y)\partial_{y^k}\delta(y-x) - \Phi^k_5(x)\partial_{x^k}\delta(y-x) \approx 0, \tag{1.97}$$

$$\{\Phi_4(y), \Phi_{5j}(x)\} = \Phi_4(x)\partial_{y^j}\delta(y-x) \approx 0, \tag{1.98}$$

$$\{\Phi_{5i}(y), \Phi_{5j}(x)\} = \Phi_{5i}(x)\partial_{y^j}\delta(y-x) - \Phi_{5j}(y)\partial_{x^i}\delta(y-x) \approx 0. \tag{1.99}$$

The set $\{\Phi_1, \Phi^i_2, \Phi_4, \Phi^i_5\}$ is complete. Namely, the time flows of the secondary constraints do not give rise to any new constraints:

$$\begin{aligned}\dot{\Phi}_4(x) &= \int d^3y\{\mathscr{H}^{(0)}_{\mathrm{ADM}}(y), \Phi_4(x)\} + \int d^3y\{\Phi_1(y), \Phi_4(x)\}\lambda_1 \\ &\quad + \int d^3y\{\Phi^i_2(y), \Phi_4(x)\}\lambda_{2i} \approx 0, \end{aligned} \tag{1.100}$$

[6] The subscripts of constraints may seem to be weird, say why does not Φ_3 exist? This convention has been made for the later convenience.

$$\dot{\Phi}_{5j}(x) = \int d^3y\{\mathscr{H}^{(0)}_{\rm ADM}(y), \Phi_{5j}(x)\} + \int d^3y\{\Phi_1(y), \Phi_{5j}(x)\}\lambda_1$$
$$+ \int d^3y\{\Phi_2^i(y), \tilde{\Phi}_{5j}(x)\}\lambda_{2i} \approx 0. \tag{1.101}$$

All Poisson brackets among the constraints yield linear combinations of constraints. Thus, following Dirac's theory of constrained systems, one finds

$$\#(\text{phase-space variables}) = \#(h_{ij}, p^{ij}, N, p_N, N_i, p^i_{\mathbf{N}}) = 2(6+1+3) = 20,$$
$$\#(\text{2nd-class constraints}) = 0,$$
$$\#(\text{1st-class constraints}) = \#(\Phi_1, \Phi_{2j}, \Phi_4, \Phi_{5j}) = 1+3+1+3 = 8.$$

Consequently, the number of physical degrees of freedom (DOF) in general gravity reads

$$\text{DOF} = \frac{1}{2}(20 - 0 - 2\times 8) = 2. \tag{1.102}$$

DOF coincides with that of the spin-2 graviton. Following general nomenclatures, we call Φ_4 and Φ_5^i, *Hamiltonian constraint* and *momentum constraint*, respectively.

In the following, we will check the relation between first-class constraints and gauge transformations. We introduce smooth test vector fields, $\xi^\mu = (\xi^0, \xi^i)$ and $\eta^\mu = (\eta^0, \eta^i)$, which fall off fast enough to suppress all the boundary contributions [37]. Henceforth, we define the smeared constraints:

$$\hat{\Phi}_\Sigma(\xi^0) = \int d^3x \xi^0(x)\Phi_4(x), \quad \hat{\Phi}_h(\xi^i) = \int d^3x \xi^i(x)\Phi_{5i}(x). \tag{1.103}$$

The generator $\mathscr{G}(\xi^i)$ of the spatial diffeomorphism acts on a phase-space variable A as

$$\{A(y), \mathscr{G}(\xi^i)\} = \pounds_\xi A(y). \tag{1.104}$$

Considering the reduced set of phase-space variables, (h_{ij}, p^{ij}), by solving the primary constraints, Φ_1 and Φ_2^j, one finds that the spatial diffeomorphisms for this set are generated by $(\hat{\Phi}_h, -\hat{\Phi}_h)$. To see the effect of $\hat{\Phi}_\Sigma$, we consider the following bracket:

$$\{h_{ij}(y), \hat{\Phi}_\Sigma(\xi^0)\} = \xi^0 \pounds_n h_{ij}(y). \tag{1.105}$$

From this expression, one can understand that $\hat{\Phi}_\Sigma(\xi^0)$ deform the foliation Σ toward its orthogonal direction n^μ with degree ξ^0. Thus, $\hat{\Phi}_\Sigma(\xi^0)$ is the generator of the hypersurface (foliation) deformation. The Poisson bracket among two $\hat{\Phi}_\Sigma$'s gives us more concrete understanding for the surface deformation:

$$\{\hat{\Phi}_\Sigma(\xi^0), \hat{\Phi}_\Sigma(\eta^0)\} = \hat{\Phi}_h\left(h^{ij}[\xi^0\partial_j\eta^0 - \eta^0\partial_j\xi^0]\right) \equiv \hat{\Phi}_h(S^i). \tag{1.106}$$

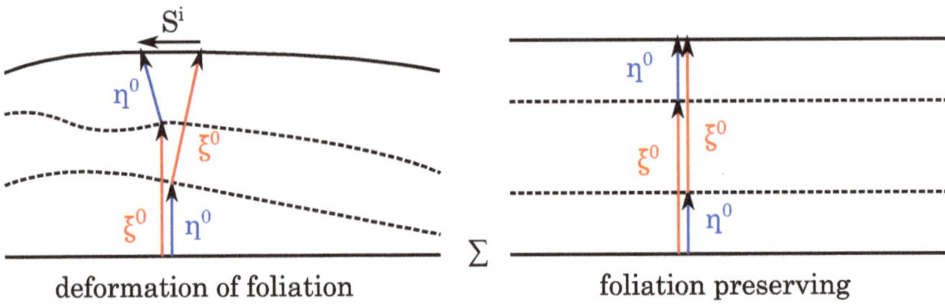

Fig. 1.10 Hypersurface (foliation) deformation

Figure 1.10 is the graphical expression of (1.106) [38]. As is clear from this figure, if the ξ^0 is a function of x^i, the initial foliation Σ can be deformed by the $\hat{\Phi}_\Sigma(\xi^0)$. Conversely, if the ξ^0 is a constant on the foliation, the foliation is unchanged by the $\hat{\Phi}_\Sigma(\xi^0)$ (see Fig. 1.10). As a consequence, we understand the following: the $\hat{\Phi}_h$ changes the intrinsic quantities, h_{ij} and p^{ij}. On the other hand, the $\hat{\Phi}_\Sigma$ deforms the extrinsic quantity, foliation. Thanks to the $\hat{\Phi}_\Sigma$, the freedom of how to choose the time axis is translated into gauge degrees of freedom, hypersurface deformation.

1.3.2 Hořava-Lifshitz Gravity

Hořava proposed a quantum theory of gravity at a Lifshitz point called *Hořava-Lifshitz gravity* [39]. Expecting an ultraviolet fixed point where space and time scale in a different manner such that

$$x^i \to bx^i, \ t \to b^z t, \tag{1.107}$$

he tried to make gravity power-counting renormalizable so as to avoid the unitarity violation. The exponent z in (1.107) is called *dynamical critical exponent*. Setting z large enough, one can obtain the gravitational coupling constant with non-negative mass dimensions, i.e., power-counting renormalizable. The unitarity can be preserved at least perturbatively if one keeps time and space derivatives are of first order and of z-th order at most in the action, respectively: the re-summed graviton propagator includes no ghost excitations. Since Hořava-Lifshitz (HL) gravity has the built-in asymmetry between space and time, it fits well with the ADM-like variables and so with the foliation. Thus, the action of HL gravity in 3 + 1 dimensions is given by

$$S_{\mathrm{HL}} = -\frac{1}{\kappa}\int dt d^3x \sqrt{h} N \left(K_{ij}K^{ij} - \lambda K^2 - \mathscr{V}[h_{ij}]\right), \tag{1.108}$$

where κ and λ are the gravitational coupling constant analogous to $16\pi G_4$ in general relativity and a dimensionless constant, respectively; h_{ij} is the spatial metric on the foliation; K_{ij} is the extrinsic curvature defined by the "lapse function" N and the "shift vector" N^i as

$$K_{ij} = \frac{1}{2N}(\dot{h}_{ij} - \nabla_i N_j - \nabla_j N_i). \tag{1.109}$$

In this regard, the dot means the time derivative and ∇_i is the covariant derivative associated with the h_{ij}. We call $\mathcal{V}[h_{ij}]$ *potential* because it is made of h_{ij} and spatial derivative terms and is invariant under the spatial diffeomorphism. Notice that if one sets the mass dimension of space such that $[x^i] = -1$, the mass dimension of the gravitational coupling constant is

$$[\kappa] = z - 3. \tag{1.110}$$

Therefore, if $z \geq 3$, the action (1.108) is power-counting (super) renormalizable. In $z = 3$, ingredients of the potential are

$$\mathcal{V}[h_{ij}] = \sigma + \zeta R + \alpha R^2 + \beta R^{ij} R_{ij} + \gamma R \Delta R + \cdots, \tag{1.111}$$

where R is the 3-dimensional Ricci scalar; σ, ζ, α, β and γ are a constant with mass dimension 6, 4, 2 and 0, respectively; Δ is the Laplacian: $\Delta = h^{ij}\nabla_i \nabla_j$. According to the Wilsonian renormalization group, one should include all possible terms made of h_{ij} compatible with the spatial diffeomorphism into the potential. The symmetry of the HL action (1.108) is the foliation preserving diffeomorphism:

$$x^i \to x^i + \xi^i(t, x^i), \qquad t \to t + \xi^0(t). \tag{1.112}$$

The original motivation of HL gravity is to anticipate the action (1.108) effectively reduces to the Einstein gravity at low energies:

$$S_{\mathrm{HL}} \to -\frac{1}{\kappa}\int dt d^3x \sqrt{h} N \left(K_{ij}K^{ij} - \lambda K^2 - \sigma - \zeta R\right), \tag{1.113}$$

where λ is expected to approach 1 along the renormalization group flow: the full diffeomorphism can be approximately recovered at low energies. However, it is unclear if the parameter λ really flows to 1. Besides, in HL gravity the full diffeomorphism is explicitly broken by the anisotropic scaling, so that the extra digrees of freedom (DOF) in addition to DOF of the traceless and transverse mode, i.e., spin-2 graviton. Thus, one should care about behaviors of the extra mode at low energies. Before explaining behaviors of the extra mode, we clarify how it appears based on the Hamiltonian formalism. In passing to the Hamiltonian formalism, we set the notation for the canonical conjugate momenta as

$$L_{\mathrm{HL}}\left((h_{ij}, \dot{h}_{ij}), (N, \dot{N}), (N_i, \dot{N}_i)\right) \to H_{\mathrm{HL}}\left((h_{ij}, p^{ij}), (N, p_N), (N_i, p_{\mathrm{N}}^i)\right), \tag{1.114}$$

where $S_{\mathrm{HL}} = \int dt L_{\mathrm{HL}}$. As in general relativity, the time derivatives of N and N_i are absent. Thus, we have the primary constraints,

$$\Phi_1 \equiv p_N = 0, \quad \Phi_2^i \equiv p_{\mathbf{N}}^i = 0. \tag{1.115}$$

Imposing the consistency of the primary constraints under the time flows, one finds

$$[\text{descendants of} \, p_N = 0] \qquad \Phi_1 = 0 \rightarrow \dot{\Phi}_1 = \Phi_4 \approx 0 \rightarrow \dot{\Phi}_4 = \Phi_8 \approx 0; \tag{1.116}$$
$$[\text{descendants of} \, p_{\mathbf{N}} = 0] \qquad \Phi_2^i = 0 \rightarrow \dot{\Phi}_2^i = \Phi_5^i \approx 0. \tag{1.117}$$

One finds that the set, $\{\Phi_1, \Phi_2^i, \Phi_4, \Phi_5^i, \Phi_8\}$, is complete. The classification of each constraint and the number of constraints and phase-space variables can be found:

$$\begin{aligned}
&\#(\text{phase-space variables}) = \#(h_{ij}, p^{ij}, N, p_N, N_i, p_{\mathbf{N}}^i) = 2(6+1+3) = 20,\\
&\#(\text{2nd-class constraints}) = \#(\Phi_1, \Phi_4, \Phi_8) = 1+1+1 = 3,\\
&\#(\text{1st-class constraints}) = \#(\Phi_{2j}, \Phi_{5j}) = 3+3 = 6.
\end{aligned}$$

Consequently, DOF of HL gravity becomes [40]

$$\text{DOF} = \frac{1}{2}(20 - 3 - 2 \times 6) = 2 + \frac{1}{2}. \tag{1.118}$$

Thus, the extra DOF is 1/2. Getting one half DOF might not necessarily be problematic in itself.[7] There are, however, at least three problematic features related to the existence of this new half degree of freedom in HL gravity. The first problem concerns the absence of dynamics and is associated to the fact that the extra DOF is halved. For generic asymptotically flat space-times, the lapse is forced to be zero at spatial infinity [40]. Actually, for particular values of the couplings, it was shown (and suggested this could be the case for generic values of the couplings) that the lapse must vanish everywhere. This indicates that there is no dynamics in HL gravity. The second problem is the short distance instability the extra mode might trigger (if the dynamics was not frozen). Looking at perturbations around generic backgrounds, it was found in [41] that the high frequency modes of the extra DOF develop an imaginary part and the perturbations can grow very swiftly in time.[8] The third problem is the self-coupling of the scalar mode which remains strong to very low energy scales [41, 42]. This implies that the extra mode never decouples and thus HL gravity does not flow to general relativity in the infrared as it was hoped.

To avoid the pathologies related to the one half DOF, Blas, Pujolas and Sibiryakov proposed the so-called *healthy extension* of HL gravity [43]. To make HL gravity "healthy", they introduced the following 3-vector:

[7] A well-known example is a chiral boson.

[8] The first two issues are apparently contradictory: if the lapse must collapse everywhere, there would be no way to develop an exponentially growing mode involving the lapse. We will, however, discuss the instabilities of the n-DBI model which evades the first issue.

$$a_i = \frac{\partial_i N}{N}; \tag{1.119}$$

they added gauge (foliation preserving diffeomorphism) invariant quantities constructed by a_i's into the potential. This non-linear lapse dependence cures the pathologies. Especially, we will explain how the most serious problem, i.e., the vanishing "lapse" problem, can be cured in the healthy extension. Because of the lapse dependence from a_i, the time-flow of the Hamiltonian constraint Φ_4 does not create further constraint and it simply determines the Lagrange multiplier of Φ_1. In the original HL gravity, the constraint equation, $\Phi_8 \approx 0$, is linear in the lapse; always one obtains $N = 0$ as the solution. In fact, this is the origin of the vanishing lapse problem. The healthy extension of HL gravity succeeded in evading the issue. As shown in [8], the model called n-DBI gravity naturally includes the non-linear lapse dependence, which makes n-DBI gravity "healthier" than the original HL gravity at least on this point.

1.3.3 n-DBI Gravity

According to the Big Bang theory [44, 45], the Universe started from enormously hot and dense state. It was about 13.75 billion years ago. Around 10^{-43} seconds after the birth, called *Planck time*, gravitational force began to be week and the Universe was filled with scalar fields called *inflatons*. At the Planck time or more later, say 10^{-35} seconds, i.e., the scale of grand unified theory (GUT), the Universe went through an exponential growth driven by the inflatons. This scenario is called *inflation* [46–52]. Lots of observations of the cosmic microwave background radiation (CMBR) suggest the approximate scale invariance of spatial variations in energy at the early stage of the Universe. The inflation scenario naturally explains this nearly scale-invariant spectrum. Searching for candidates of the inflatons from would-be fundamental theories is a natural direction one should take. However, so far no one could single out scalar fields as the inflatons from a bunch of candidates.

Herdeiro and Hirano proposed an alternation of Einstein's gravity using the scale invariance as a guiding principle [6]. It is dubbed as n-DBI gravity. This model was designed to yield non-eternal inflation spontaneously without introducing any scalar fields. n-DBI gravity was named after two characteristic features of the model: first, it becomes a Dirac-Born-Infeld (DBI) type conformal scalar theory when the Universe is conformally flat and a conformal mode of the metric plays the role of the scalar field agent of inflation; second, it contains the space-time foliation provided by an everywhere time-like vector field n, which couples to the gravitational sector of the theory, but decouples in the small curvature limit. The scale invariance is, indeed, a nice guiding principle to investigate cosmological issues. Before explaining n-DBI gravity, we will show how the scale invariance works for cosmological issues; although the argument is not satisfactory, it may cultivate a sence of what we mean. We start with the Einstein-Hilbert action with a cosmological constant term:

$$S = -\frac{1}{16\pi G_4} \int_{\mathcal{M}} d^4x \sqrt{-g}(R - 2\Lambda), \tag{1.120}$$

where G_4 and Λ are the Newton constant and the cosmological constant, respectively. We consider the conformally flat metric ansatz:

$$g_{\mu\nu} = l_p^2 \phi(t, x^i)^2 \eta_{\mu\nu}, \tag{1.121}$$

where ϕ is a conformal mode, $l_p(= \sqrt{G_4})$ is the Planck length and $\eta_{\mu\nu}$ is the flat Minkowski metric. Plugging this conformally flat metric into (1.120), one finds the classically conformal action:

$$S = \frac{3}{4\pi} \int d^4x \left[-\frac{1}{2} \eta^{\mu\nu} \partial_\mu \phi \partial_\nu \phi + \frac{1}{6} G_4 \Lambda \phi^4 \right], \tag{1.122}$$

which holds up to total derivative terms. After a suitable rotation, this action can be seen as the $\lambda\phi^4$ theory where $\lambda = G_4\Lambda$. Because of the infrared triviality of the $\lambda\phi^4$ theory, the cosmological constant reduces to 0 along the renormalization group flow [53]. We stress that the $\lambda\phi^4$ theory is classically conformal and therefore scale invariant. Thus, the scale invariance answers why the cosmological constant is so small to some extent. It has been known that the conformal scalar theory has a unique generalization [54]:

$$S_{\text{CFT}} = -T_{\text{D3}} \int d^4x \lambda\phi^4 \left(\sqrt{1 + \frac{\eta^{\mu\nu} \partial_\mu \phi \partial_\nu \phi}{\lambda\phi^4}} - q \right), \tag{1.123}$$

where λ and q are arbitrary dimension-less constants can not be fixed by conformal invariance. When considering the type IIB superstring theory on $\text{AdS}_5 \times S^5$, this is the DBI action of a single D3-brane, with tension T_{D3}, stuck in $\text{AdS}_5 \times S^5$ where ϕ is the AdS radial coordinate in the Poincaré patch. If ϕ is large, then S_{CFT} reduces to the $\lambda\phi^4$ theory. One can ask, "What is the gravitational theory reducing to the S_{CFT} by a suitable metric ansatz, if it exists?" It indeed exists: n-DBI gravity. Here we give the action of n-DBI gravity without matters [6, 7]:

$$S_{\text{nDBI}} = -\frac{3\lambda}{4\pi G_4^2} \int_{\mathcal{M}} d^4x \sqrt{-g} \left\{ \sqrt{1 + \frac{G_4}{6\lambda} \left({}^{(4)}R + \mathcal{K} \right)} - q \right\}, \tag{1.124}$$

where G_4 is the Newton constant; q and λ are the two dimensionless parameters of the theory; ${}^{(4)}R$ is the four dimensional Ricci scalar. To completely define the theory a foliation structure must be chosen. Let n be a unit time-like vector field, everywhere orthogonal to the leaves of such foliation; we set that $h_{\mu\nu} = g_{\mu\nu} + n_\mu n_\nu$ and $K_{\mu\nu} = \frac{1}{2}\pounds_n h_{\mu\nu}$. Then $\mathcal{K}$ in (1.124) has been defined as

$$\mathscr{K} \equiv -\frac{2}{\sqrt{h}}\pounds_n(\sqrt{h}K), \qquad K \equiv K_{\mu\nu}h^{\mu\nu}. \tag{1.125}$$

This scalar quantity is closely related to the Gibbons-Hawking-York boundary term [55, 56]. However, since $\mathscr{K}$ is in the square root, it can not become a boundary term: the vector field n couples to the bulk gravity via $\mathscr{K}$. Because of this fact, the full diffeomorphism is broken in n-DBI gravity. When plugging conformally flat and homogeneous metric ansatz, $g_{\mu\nu} = l_p^2\phi(t)^2\eta_{\mu\nu}$, into the action of n-DBI gravity (1.124), one can obtain the DBI scalar action (1.123) up to an overall constant. The role of $\mathscr{K}$ is to convert the second-order time derivative of ϕ in ${}^{(4)}R$ into the first-order derivative. Namely,

$$^{(4)}R = \frac{6}{l_p^2}\cdot\frac{\ddot{\phi}}{\phi^3}, \qquad ^{(4)}R + \mathscr{K} = -\frac{6}{l_p^2}\cdot\frac{\dot{\phi}^2}{\phi^4}, \tag{1.126}$$

where we have plugged the conformally flat and homogeneous metric. Thereby, the anisotropy among space and time shows up. Therefore, as claimed before n-DBI gravity does not possess the full diffeomorphism. The gauge symmetry of n-DBI gravity is the so-called *foliation preserving diffeomorphism*:

$$t \to t + \xi^0(t), \;\; x^i \to x^i + \xi^i(t, x^i), \tag{1.127}$$

where $\xi^0(t)$ and $\xi^i(t, x^i)$ are infinitesimal and arbitrary. Since ξ^0 depends only on t, the form of foliation can not be changed. One may ask, "Do you have any plausible reasons to break the space-time gauge symmetry?" The answer is yes. "*There are reasons to suspect that the vacuum in quantum gravity may determine a preferred rest frame at the microscopic level. However, if such a frame exists, it must be very effectively concealed from view* (by Jacobson in the paper about Einstein-æther gravity [5])." However, as mentioned by Jacobson implicitly, in the infrared regime, the full diffeomorphism ought to be recovered to a very good accuracy. n-DBI gravity can pass this condition naturally: expanding the action (1.124) by small G_4, one can obtain the Einstein-Hilbert action with a cosmological constant term and the Gibbons-Howking-York bounday term at the lowest level. Alternatively, one can consider the following conventional limit:

$$\lambda \to \infty, \;\; q \to 1, \tag{1.128}$$

under the quantity $\lambda(1-q)$ is fixed [6, 7]. This limit is called *Einstein gravity limit* because one succeeds in letting n decoupled completely and obtains

$$S_{\text{nDBI}} \to S = -\frac{1}{16\pi G_4}\int_{\mathscr{M}} d^4x\sqrt{-g}(^{(4)}R - 2\Lambda_{\text{Einstein}}) + \frac{1}{8\pi G_4}\int_{\partial\mathscr{M}} d^3x\sqrt{h}K, \tag{1.129}$$

where

$$\Lambda_{\rm Einstein} = \frac{6\lambda(q-1)}{G_4}. \tag{1.130}$$

For later convenience, we will show alternative expressions of n-DBI gravity below [8].

1.3.3.1 ADM Form

Performing the Arnowitt-Deser-Misner (ADM) decomposition [35, 36]

$$ds^2 = -N^2 dt^2 + h_{ij}\left(dx^i + N^i dt\right)\left(dx^j + N^j dt\right), \tag{1.131}$$

where N and N_i, respectively, are the lapse and shift functions, and h_{ij} with $i, j = 1, 2, 3$ is the spatial metric, the action (1.124) becomes

$$S = -\frac{3\lambda}{4\pi G_4^2}\int dt d^3x\sqrt{h}N\left[\sqrt{1+\frac{G_4}{6\lambda}\left(R + K_{ij}K^{ij} - K^2 - 2N^{-1}\Delta N\right)} - q\right], \tag{1.132}$$

where R is the Ricci scalar of h_{ij}. In particular, observe the lapse term $N^{-1}\Delta N$. As suggested in [43], adding analogous terms can provide a consistent extension of Hořava-Lifshitz (HL) gravity. Indeed, as we shall see later, this term plays a crucial role in evading the pathologies that afflicted HL gravity.

1.3.3.2 Linearlized Form with Auxiliary Field

As it is known in various contexts, square root type actions may be linearized by introducing auxiliary fields. A well known example is the classical equivalence between the Polyakov and Nambu-Goto actions in string theory, via the introduction of an auxiliary metric (the world sheet metric). A similar reformulation for the Eddington inspired Born-Infeld gravity was provided recently [57], for which the auxiliary variable is the "apparent" metric. In our case, with the introduction of an auxiliary scalar field e, the action (1.124) becomes[9]

$$S^e = -\frac{1}{16\pi G_4}\int d^4x\sqrt{-g}e\left[{}^{(4)}R - 2\Lambda_C(e) + \mathcal{K}\right], \quad \Lambda_C(e) = \frac{3\lambda}{G_4}\left(\frac{2q}{e} - 1 - \frac{1}{e^2}\right). \tag{1.133}$$

For constant field e, this is the Einstein-Hilbert action with a cosmological constant (in exactly the form found in [7] in terms of the integration constant C therein) with the Gibbons-Hawking-York boundary term [55, 56]. For generic e, the theory resembles general relativity in a Jordan frame (but indeed it is quite different).

---

[9] We thank Soo-Jong Rey and Takao Suyama for suggesting this formulation.

It is worth noting that we can rewrite the action (1.133) in the Einstein frame by performing a Weyl transformation:

$$g_{\mu\nu} \rightarrow e^{-1} g_{\mu\nu}. \tag{1.134}$$

Redefining the auxiliary field $e \equiv \exp(2\chi)$, the action becomes, up to boundary terms,

$$S_{\text{Einstein}} = -\frac{1}{16\pi G_4} \int d^4x \sqrt{-g} \left[{}^{(4)}\!R - 6\left(n^\alpha n^\beta + h^{\alpha\beta}\right) \partial_\alpha \chi \partial_\beta \chi + 2\mathscr{K}\chi + V(\chi) \right], \tag{1.135}$$

with the potential

$$V(\chi) = \frac{6\lambda}{G_4} \exp(-4\chi) \left[(\exp(\chi) - \exp(-\chi))^2 + 2(1-q) \right]. \tag{1.136}$$

There is a caveat: it is misleading to regard the Einstein frame theory as a scalar-tensor theory (the scalar being χ). Despite the appearance of the kinetic term, the scalar field χ is still an auxiliary field and does not give rise to an *independent* degree of freedom. As we will see, the extra scalar mode is furnished in the metric and the scalar field χ is only related to it through the equations of motion.

1.3.3.3 Covariant Form with Stückelberg Field

n-DBI gravity breaks Lorentz invariance due to the coupling of gravity to the unit time-like vector field n, which defines a preferred space-time foliation. Full general covariance can be restored by introducing a Stückelberg field $\phi(x^\mu)$, such that its gradient is everywhere time-like and non-vanishing:

$$n_\mu = -\frac{\partial_\mu \phi}{\sqrt{-X}}, \qquad g^{\mu\nu} n_\mu n_\nu = -1, \qquad X \equiv g^{\mu\nu} \partial_\mu \phi \partial_\nu \phi. \tag{1.137}$$

By definition the theory is invariant under $\phi \rightarrow f(\phi)$, where $f(\phi)$ is an arbitrary function of ϕ. Note that the original non-covariant form is recovered when $\phi = t$ for which $n = (-N, 0, 0, 0)$; the symmetry reduces to the time reparametrization $t \rightarrow f(t)$. A similar treatment in the case of HL gravity has been performed in [41].

The extrinsic curvature becomes

$$K_{\mu\nu} = \frac{1}{2} \left[n^\alpha D_\alpha h_{\mu\nu} + h_{\mu\alpha} D_\nu n^\alpha + h_{\nu\alpha} D_\mu n^\alpha \right], \qquad K = D_\alpha n^\alpha. \tag{1.138}$$

It then follows that

$$\mathscr{K} = -2 D_\alpha \left(n^\alpha D_\beta n^\beta \right). \tag{1.139}$$

Thus, the n-DBI action may be rewritten in the covariant form

$$S_c = -\frac{3\lambda}{4\pi G_4^2}\int d^4x\sqrt{-g}\left\{\sqrt{1+\frac{G_4}{6\lambda}\left({}^{(4)}R - 2D_\alpha\left[\frac{\partial^\alpha\phi}{\sqrt{-X}}D_\beta\left(\frac{\partial^\beta\phi}{\sqrt{-X}}\right)\right]\right)} - q\right\}. \tag{1.140}$$

In contrast to the Einstein frame theory, this covariant theory can be thought of as a scalar-tensor theory. Indeed, in this form, the scalar field ϕ does yield an independent degree of freedom. Put differently, the scalar mode in the metric is entirely transferred to the Stückelberg field ϕ.

Clearly, this can be linearized again by introducing the auxiliary field e,

$$S_c^e = -\frac{1}{16\pi G_4}\int d^4x\sqrt{-g}e\left\{{}^{(4)}R - 2\Lambda_C(e) - 2D_\alpha\left[\frac{\partial^\alpha\phi}{\sqrt{-X}}D_\beta\left(\frac{\partial^\beta\phi}{\sqrt{-X}}\right)\right]\right\}. \tag{1.141}$$

In these formulations, general covariance gets *spontaneously* broken by the scalar field ϕ acquiring the vev $\langle\phi\rangle = t$. It should be noted that this has a certain bearing on the ghost condensation of [58].

References

1. Teitelboim, C. (1983). Causality versus gauge invariance in quantum gravity and supergravity. *Physical Review Letters*, *50*, 705.
2. Wheeler, J. A. (1970). In R. Gilbert, & R. Newton (Eds.) *Analytical methods in mathematical physics*. New York: Gordon and Breach.
3. Ambjorn, J., & Loll, R. (1998). Non-perturbative Lorentzian quantum gravity, causality and topology change. *Nuclear Physics B*, *536*, 407. [hep-th/9805108].
4. Ambjorn, J., Gorlich, A., Jurkiewicz, J., & Loll, R. (2008). Planckian birth of the quantum de sitter universe. *Physical Review Letters*, *100*, 091304. arXiv:0712.2485 [hep-th].
5. Jacobson, T. Einstein-aether gravity: A Status report, PoS QG -PH (2007) 020. arXiv:0801.1547 [gr-qc].
6. Herdeiro, C., & Hirano, S. (2012). Scale invariance and a gravitational model with non-eternal inflation. *Journal of Cosmology and Astroparticle Physics*, *1205*, 031. arXiv:1109.1468 [hep-th].
7. Herdeiro, C., Hirano, S., & Sato, Y. (2011). n-DBI gravity. *Physical Review D*, *84*, 124048. arXiv:1110.0832 [gr-qc].
8. Coelho, F. S., Herdeiro, C., Hirano, S., & Sato, Y. (2012). On the scalar graviton in n-DBI gravity. *Physics Letters B*, *86*, 064009. arXiv:1205.6850 [hep-th].
9. Ambjorn, J., Goerlich, A., Jurkiewicz, J., & Loll, R. (2012). Nonperturbative Quantum Gravity. *Physics Reports*, *519*, 127. arXiv:1203.3591 [hep-th].
10. Regge, T. (1961). General relativity without coordinates. *Nuovo Cimento Series*, *19*, 558.
11. Weyl, H. (1922). *Space, time, matter*. London: Methuen.
12. Codello, A., Percacci, R., & Rahmede, C. (2009). Investigating the ultraviolet properties of gravity with a Wilsonian renormalization group equation. *Annals of Physics*, *324*, 414. arXiv:0805.2909 [hep-th].
13. Reuter, M., & Saueressig, F. (2007). Functional Renormalization Group Equations, Asymptotic Safety, and Quantum Einstein Gravity. arXiv:0708.1317 [hep-th] (Based on lectures given by

M.R. at the "First Quantum Geometry and Quantum Gravity School", Zakopane, Poland, March 2007, and the "Summer School on Geometric and Topological Methods for Quantum Field Theory", Villa de Leyva, Colombia, July 2007, and by F.S. at NIKHEF, Amsterdam, The Netherlands, June 2006).
14. Niedermaier, M., & Reuter, M. (2006). The asymptotic safety scenario in quantum gravity. *Living Reviews in Relativity*, *9*, 5.
15. Hamber, H. W., & Williams, R. M. (2005). Nonlocal effective gravitational field equations and the running of Newton's G. *Physical Review D*, *72*, 044026. [hep-th/0507017].
16. Christiansen, N., Litim, D. F., Pawlowski, J. M., & Rodigast, A. (2014). Fixed points and infrared completion of quantum gravity. *Physics Letters B*, *728*, 114. arXiv:1209.4038 [hep-th].
17. Weinberg, S. (1979). Ultraviolet divergences in quantum theories of gravitation. In S. W. Hawking, W. Israel (Eds.) *General relativity: Einstein centenary survey* (pp. 790–831) Cambridge: Cambridge University Press.
18. Reuter, M. (1998). Nonperturbative evolution equation for quantum gravity. *Physical Review D*, *57*, 971. [hep-th/9605030].
19. Ponzano, G., & Regge, T. (1968). In F. Bloch (Ed.) *Spectroscopic and group theoretical methods in physics* Amsterdam: North-Holland.
20. Turaev, V. G., & Viro, O. Y. (1992). State sum invariants of 3 manifolds and quantum 6j symbols. *Topology*, *31*, 865.
21. Ambjorn, J., Durhuus, B., & Frohlich, J. (1985). Diseases of triangulated random surface models, and possible cures. *Nuclear Physics B*, *257*, 433.
22. Ambjorn, J., Durhuus, B., Frohlich, J., & Orland, P. (1986). The appearance of critical dimensions in regulated string theories. *Nuclear Physics B*, *270*, 457.
23. David, F. (1985). Planar diagrams, two-dimensional lattice gravity and surface models. *Nuclear Physics B*, *257*, 45.
24. Billoire, A., & David, F. (1986). Microcanonical simulations of randomly triangulated planar random surfaces. *Physics Letters B*, *168*, 279.
25. Kazakov, V. A., Migdal, A. A., & Kostov, I. K. (1985). Critical properties of randomly triangulated planar random surfaces. *Physics Letters B*, *157*, 295.
26. Boulatov, D. V., Kazakov, V. A., Kostov, I. K., & Migdal, A. A. (1986). Analytical and numerical study of the model of dynamically triangulated random surfaces. *Nuclear Physics B*, *275*, 641.
27. Ambjorn, J., Jurkiewicz, J., & Loll, R. (2001). Dynamically triangulating Lorentzian quantum gravity. *Nuclear Physics B*, *610*, 347. [hep-th/0105267].
28. Ambjorn, J., Loll, R., Watabiki, Y., Westra, W., & Zohren, S. (2008). Topology change in causal quantum gravity. arXiv:0802.0896 [hep-th] (proceedings of the workshop JGRG 17 (Nagoya, Japan, December 2007)).
29. Ambjorn, J., Gorlich, A., Jurkiewicz, J. & Loll, R. (2010). CDT—an entropic theory of quantum gravity. arXiv:1007.2560 [hep-th] (Lectures presented at the "School on Non-Perturbative Methods in Quantum Field Theory" and the "Workshopon Continuum and Lattice Approaches to Quantum Gravity", Sussex, September 15th–19th 2008. Toappear as a contribution to a Springer Lecture Notes in Physics book).
30. Ambjorn, J., Chekhov, L., Kristjansen, C. F., & Makeenko, Y. (1993). Matrix model calculations beyond the spherical limit. *Nuclear Physics B*, *404*, 127 [Erratum-ibid. B 449 (1995) 681] [hep-th/9302014].
31. Ambjorn, J., Loll, R., Watabiki, Y., Westra, W., & Zohren, S. (2008). A new continuum limit of matrix models. *Physics Letters B*, *670*, 224. arXiv:0810.2408 [hep-th].
32. Polyakov, A. M. (1981). Quantum geometry of bosonic strings. *Physics Letters B*, *103*, 207.
33. Nakayama, Y. (2004). Liouville field theory: A decade after the revolution. *International Journal of Modern Physics A*, *19*, 2771. [hep-th/0402009].
34. Ambjorn, J., Loll, R., Watabiki, Y., Westra, W., & Zohren, S. (2008). A matrix model for 2D quantum gravity defined by causal dynamical triangulations. *Physics Letters B*, *665*, 252. arXiv:0804.0252 [hep-th].

35. Arnowitt, R. L., Deser, S., & Misner, C. W. (1960). Canonical variables for general relativity. *Physical Review*, *117*, 1595.
36. Arnowitt, R. L., Deser, S., & Misner, C. W. (2008). The dynamics of general relativity. *General Relativity and Gravitation*, *40*, 1997–2027. arXiv:gr-qc/0405109.
37. Khoury, J., Miller, G. E. J. & Tolley, A. J. (2012). Spatially covariant theories of a transverse, traceless graviton, part I: Formalism. *Physical Review D*, *85*, 084002. arXiv:1108.1397 [hep-th].
38. Bojowald, M. (2010). *Canonical gravity and applications*. Cambridge: Cambridge University Press.
39. Horava, P. (2009). Quantum gravity at a Lifshitz point. *Physical Review D*, *79*, 084008. arXiv:0901.3775 [hep-th].
40. Henneaux, M., Kleinschmidt, A., & Gomez, G. L. (2010). A dynamical inconsistency of Horava gravity. *Physical Review D*, *81*, 064002. arXiv:0912.0399 [hep-th].
41. Blas, D., Pujolas, O., & Sibiryakov, S. (2009). On the extra mode and inconsistency of Horava gravity. *Journal of High Energy Physics*, *0910*, 029. arXiv:0906.3046 [hep-th].
42. Charmousis, C., Niz, G., Padilla, A., & Saffin, P. M. (2009). Strong coupling in Horava gravity. *Journal of High Energy Physics*, *0908*, 070. arXiv:0905.2579 [hep-th].
43. Blas, D., Pujolas, O., & Sibiryakov, S. (2010). Consistent extension of Horava gravity. *Physical Review Letters*, *104*, 181302. arXiv:0909.3525 [hep-th].
44. Lemaître, G. (1927). Un Univers homogeéne de masse constante et de rayon croissant rendant compte de la vitesse radiale des nbuleuses extra-galactiques. *Annales Societatis Science Bruxelles A*, *47*, 49.
45. Hubble, E. (1929). A relation between distance and radial velocity among extra-galactic nebulae. *Proceedings of the National Academy of Sciences*, *15*, 168.
46. Guth, A. H. (1981). The inflationary universe: A possible solution to the horizon and flatness problems. *Physical Review D*, *23*, 347–356.
47. Linde, A. D. (1982). A new inflationary universe scenario: A possible solution of the horizon, flatness, homogeneity, isotropy and primordial monopole problems. *Physics Letters B*, *108*, 389.
48. Linde, A. D. (1983). Chaotic inflation. *Physics Letters B*, *129*, 177.
49. Albrecht, A., & Steinhardt, P. J. (1982). Cosmology for grand unified theories with radiatively induced symmetry breaking. *Physical Review Letters*, *48*, 1220.
50. Starobinsky, A. A. (1980). A new type of isotropic cosmological models without singularity. *Physics Letters B*, *91*, 99–102.
51. Starobinsky, A. A. (1979). Relict gravitation radiation spectrum and initial state of the universe. *JETP Letters*, *30*, 682–685. (In Russian).
52. Sato, K. (1981). First order phase transition of a vacuum and expansion of the universe. *Monthly Notices of the Royal Astronomical Society*, *195*, 467.
53. Polyakov, A. M. (2006). Beyond space-time [hep-th/0602011] (based on the remarks made at the 23 Solvay Conference).
54. Maldacena, J. M. (1998). The Large N limit of superconformal field theories and supergravity. *Advanced Theories in Mathematical Physics*, *2*, 231. [hep-th/9711200].
55. Gibbons, G. W., & Hawking, S. W. (1977). Action integrals and partition functions in quantum gravity. *Physical Review D*, *15*, 2752.
56. York, J. W. (1972). Role of conformal three geometry in the dynamics of gravitation. *Physical Review Letters*, *28*, 1082.
57. Delsate, T., & Steinhoff, J. (2012). New insights on the matter-gravity coupling paradigm. *Physical Review Letters*, *109*, 021101. arXiv:1201.4989 [gr-qc].
58. Arkani-Hamed, N., Cheng, H.-C., Luty, M. A., & Mukohyama, S. (2004). Ghost condensation and a consistent infrared modification of gravity. *Journal of High Energy Physics*, *0405*, 074. [hep-th/0312099].

Chapter 2
Causal Dynamical Triangulation

2.1 Matter-Coupled CDT

In physics, when one encounters a recondite issue, it is often very effective to introduce a simple toy model. This ought to capture its nature and substance by getting rid of extraneous data. In the case of profound quantum gravity, one promising toy model is the 2-dimensional theory. A method called dynamical triangulation (DT) nicely expresses physics of the 2-dimensional Euclidean quantum gravity. Especially multicritical models of DT have been designed to describe the 2-dimensional Euclidean quantum gravity coupled to matters by Kazakov [1]. On this line of thought, Staudacher has succeeded in identifying the first multicritical point with a rational minimal conformal field theory characterized by the negative central charge, $c = -22/5$, coupled to the 2-dimensional Euclidean quantum gravity [2]. A possible way to mount the next step is to investigate a Lorentzian model for the 2-dimensional quantum gravity. One of the candidates is causal dynamical triangulation (CDT) [3]. In [4], we have proposed new multicritical models which naturally capture physics of the matter-coupled CDT in the continuum limit. In the following, we indict its essence.

2.1.1 Causal Random Geometries Coupled to Dimers

In Sect. 1.2.3, we have introduced the unrestricted triangulations as geometries discretized by any kinds of polygons. Following this thought, we introduce the potential:

$$V(z) = \frac{1}{2}z^2 - gz - \frac{g}{3}z^3 - \frac{g^3\xi}{2}z^4. \tag{2.1}$$

This potential generates geometries discretized by 1-gons, triangles and squares. Viewing each square as two triangles, one can think of the squares as part of the triangulation, but with a dimer placed on the diagonal, with a fugacity $\tilde{\xi} = g\xi$. In

Y. Sato, *Space-Time Foliation in Quantum Gravity*, Springer Theses,
DOI: 10.1007/978-4-431-54947-5_2, © Springer Japan 2014

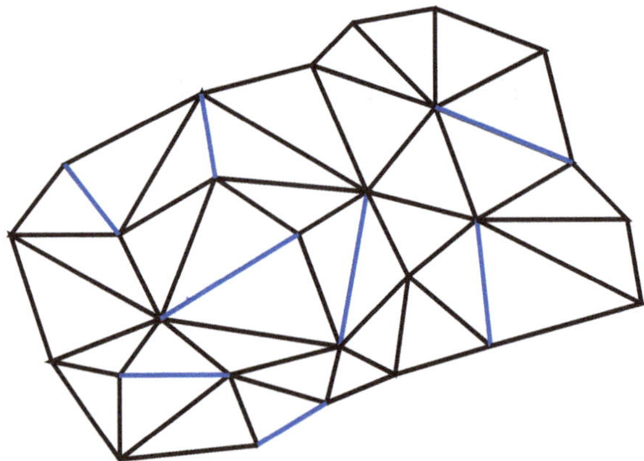

Fig. 2.1 Dimers (*blue edges*) on a triangulation

this way the model describes dimers put on random triangulations in a special way, such that there is at most one dimer per triangle. On the graph dual to the triangulation the dimers are precisely hard dimers: one dimer is allowed to be attached to each vertex at most (see Fig. 2.1). We will call them hard dimers also on the triangulation, even if the rule of putting down the dimers is slightly different from the standard hard dimer rule. Similarly we will denote ξ the fugacity of the dimers, although it is strictly speaking $\tilde{\xi}$ which serves as the fugacity. When considering the sphere topology and imposing the single-cut structure, the solution of the loop equation becomes

$$w(g,z) = \frac{1}{2g_s}\left(V'(z) - \sum_{k=1}^{3} M_k(z-a)^{k-1}\sqrt{(z-a)(z-b)}\right), \tag{2.2}$$

where $w(g,z)$ is the generating function for the boundary loop (sometimes called resolvent or disk function); g_s is the string coupling constant; (a,b) are end points of the cut. From now on we call $w(g,z)$ *resolvent*. From the asymptotic behavior of $w(g,z)$ in $|z| \gg |a-b|$, one obtains a set of equations:

$$M_3 = -2g^3\xi, \tag{2.3}$$

$$M_2 = -g + \frac{M_3}{2}(5a+b), \tag{2.4}$$

$$M_1 = 1 + \frac{M_2}{2}(3a+b) + \frac{M_3}{8}(b^2 - 10ab - 15a^2), \tag{2.5}$$

$$g_s = \frac{1}{16}\left[M_1(b-a)^2 + \frac{1}{2}M_2(b-a)^3 + \frac{5}{16}M_3(b-a)^4\right], \tag{2.6}$$

$$g = \frac{M_1}{2}(a+b) + \frac{M_2}{8}(b^2 - 6ab - 3a^2) + \frac{M_3}{16}(b^3 - 5ab^2 + 15a^2b + 5a^3). \quad (2.7)$$

Remember the moments (1.44). The moments can be written as follows:

$$\begin{aligned} M_k &= \oint_C \frac{d\omega}{2\pi i} \frac{V'(\omega)}{(\omega - a)^{k+1/2}(\omega - b)^{1/2}} \\ &= \oint_{\bar{C}} \frac{d\bar{\omega}}{2\pi i} \left[\frac{\bar{\omega}^{k-1}}{(1 - \bar{\omega}a)^{k+1/2}(1 - \bar{\omega}b)^{1/2}} \right] V'(1/\bar{\omega}), \end{aligned} \quad (2.8)$$

where $\bar{\omega} = 1/\omega$ and the path $\bar{C}$ enclosing 0 in the $\bar{\omega}$-patch. If $\xi > 0$, one can only set M_1 as zero. However, if $\xi < 0$, one can set not only M_1 but also M_2 as zero. The critical point characterized by $M_1 = M_2 = 0$ is an example of the multicritical points. Let us investigate the multicritical behavior of this model. To begin, we impose the condition for the multicritical point:

$$M_1 = M_2 = 0. \quad (2.9)$$

Plugging (2.9) into the Eqs. (2.3)–(2.7), we find

$$a = \frac{1}{6}\left[\left(-\frac{128 g_s}{5 g^3 \xi} \right)^{1/4} - \frac{1}{g^2 \xi} \right], \quad b = \frac{1}{6}\left[-5 \left(-\frac{128 g_s}{5 g^3 \xi} \right)^{1/4} - \frac{1}{g^2 \xi} \right], \quad (2.10)$$

$$b^3 - 5ab^2 + 15a^2b + 5a^3 - 8(5a+b) = 0, \quad 4 = g^3 \xi (b^2 - 10ab - 15a^2). \quad (2.11)$$

For every value of g_s except for 0, approaching the critical point defined by (2.9) yields the Liouville field theory with the central charge $c = -22/5$. What we are interested in is the “CDT” point specified with $g_s = 0$. If one can find a non-trivial scaling relation around that point outside of the universality class of the plain CDT or the Liouville theory with the negative central charge, it turns out that one has succeeded in formulating CDT coupled to dimers. It has been achieved in [4]. Imposing the condition,

$$M_1 = M_2 = g_s = 0, \quad (2.12)$$

the critical values can be obtained (see Fig. 2.2):

$$a_* = b_* = \sqrt{3}, \quad g_* = \frac{1}{a_*}, \quad \xi_* = -\frac{a_*}{6}. \quad (2.13)$$

Then we renormalize the string coupling constant as follows:

$$g_s = G_s \varepsilon^4, \quad (2.14)$$

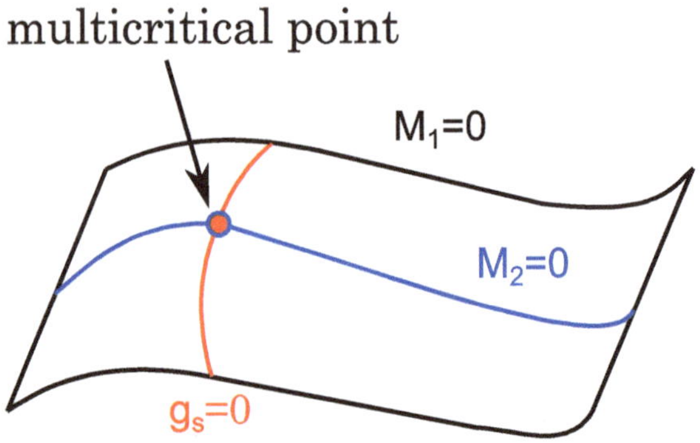

multicritical point

M$_1$=0

M$_2$=0

g$_s$=0

Fig. 2.2 Multicritical point in 3-dimensional coupling constant space (g, ξ, g_s)

where G_s is the renormalized string coupling constant and ε is the lattice spacing. Sitting on the critical line $M_1 = M_2 = 0$, one can expand g, ξ and a:

$$g_c(g_s) = g_* \left(1 - \frac{\sqrt{5}}{9} G_s^{1/2} \varepsilon^2 - \frac{16\sqrt{5}}{27} G_s^{3/4} \varepsilon^3 \right) + \mathcal{O}(\varepsilon^4), \tag{2.15}$$

$$\xi_c(g_s) = \xi_* \left(1 - \frac{\sqrt{5}}{9} G_s^{1/2} \varepsilon^2 + \frac{16\sqrt{5}}{27} G_s^{3/4} \varepsilon^3 \right) + \mathcal{O}(\varepsilon^4), \tag{2.16}$$

$$a_c(g_s) = a_* \left(1 + \frac{2}{3 \cdot 5^{1/4}} G_s^{1/4} \varepsilon \right) + \mathcal{O}(\varepsilon^2). \tag{2.17}$$

The perturbation away from $g_c(g_s)$, $\xi_c(g_s)$ which leads to the potential $V'(a, g, \xi)$ of order ε^3, assuming the boundary cosmological constant is perturbed as $z = a_c(g_s) + \varepsilon Z$, can be parametrized as

$$g = g_* + \tilde{\Lambda}\varepsilon^2 - \Lambda\,\varepsilon^3, \quad \xi = \xi_* - \frac{1}{2}\tilde{\Lambda}\varepsilon^2. \tag{2.18}$$

As in the ordinary multicritical model situation one finds a two-parameter set of solutions depending on Λ, $\tilde{\Lambda}$. Let us choose a convenient "background", using the notation from ordinary matrix models [5], which we call *CDT-background*, namely $\tilde{\Lambda} = 0$. By this choice of the background, one finds

$$V'(z; g, \xi) = \left(\Lambda + \frac{1}{9} Z^3 + \frac{1}{3}\alpha Z^2 G_s^{1/4} + \frac{1}{3}\alpha^2 Z G_s^{1/2}\right) \varepsilon^3 + G_s^{3/4} \left(\gamma + \frac{1}{9}\alpha^3\right) \varepsilon^3 + \mathcal{O}(\varepsilon^4), \tag{2.19}$$

where

$$\alpha = \frac{2}{5^{1/4}\sqrt{3}}, \quad \gamma = \frac{32\sqrt{5}}{27\sqrt{3}}. \tag{2.20}$$

When applying the CDT variables, $(Z_{\text{cdt}}, \Lambda_{\text{cdt}}, \tilde{\Lambda}_{\text{cdt}})$, defined by

$$Z = Z_{\text{cdt}} - \alpha G_s^{1/4}, \quad \Lambda = \Lambda_{\text{cdt}} - \gamma G_s^{3/4}, \tag{2.21}$$

(2.19) becomes drastically simpler:

$$V'(z; g, \xi) = \left(\Lambda_{\text{cdt}} + \frac{1}{9} Z_{\text{cdt}}^3\right) \varepsilon^3 + \mathcal{O}(\varepsilon^4). \tag{2.22}$$

One may now calculate $w(g, z)$ in the CDT limit $G_s \to 0$ where any creation of baby universes is suppressed:

$$w(g, z) = \varepsilon^{-1} W_{\text{CDT}}^{(3)}(Z_{\text{cdt}}) + \mathcal{O}(\varepsilon^0), \tag{2.23}$$

where

$$W_{\text{CDT}}^{(3)}(Z_{\text{cdt}}) = \frac{1}{Z_{\text{cdt}} + \Lambda_{\text{cdt}}^{1/3}}. \tag{2.24}$$

Next, let us take a look at observables. A leading singularity of the free energy F at the critical point provides the so-called *string susceptibility*, γ_{str}:

$$F = N^2 f_0 (g - g_*)^{2-\gamma_{\text{str}}} + \mathcal{O}(N), \tag{2.25}$$

where f_0 is some constant. The disk function (2.24) in the limit $Z_{\text{cdt}} \to \infty$ is nothing but the genus-zero free energy with a marked point: because of the limit, the boundary loop shrinks to zero yielding the sphere with an infinitesimal boundary, a marked point [see (1.34)]; to mark a point on the sphere, one needs to act the derivative with respect to g. Picking up the first non-analytic structure of the disk function in the limit $Z_{\text{cdt}} \to \infty$, one finds

$$W_{\text{CDT}}^{(3)} \sim (g - g_*)^{1-\gamma_{\text{str}}}, \tag{2.26}$$

where

$$\gamma_{\text{str}} = \frac{2}{3}. \tag{2.27}$$

In the plain CDT, $\gamma_{\text{str}} = 1/2$ [3]; one can find the same value in the so-called *branched polymer phase* of DT. On the other hand, (2.27) coincides with γ_{str} of the branched polymer coupled to dimers [4, 6]. One can compute another exponent called *edge singularity*, σ, defined as

$$\frac{d \log g_*}{d\xi} \sim (\xi - \xi_*)^{\sigma}. \tag{2.28}$$

From (2.18), one finds

$$\sigma = \frac{1}{2}. \tag{2.29}$$

This is the same as that of DT in the first multicritical point.[1]

2.1.2 Field Theory Qrising from CDT Scaling

We will show that the field-theoretic description can be obtained via the loop equation [7] (and implicitly shown in [4]). The loop equation with the potential (2.1) can be written as

$$\partial_z^2 \left(g_s w(g,z)^2 - V'(z) w(g,z) \right) = 4g^3 \xi. \tag{2.30}$$

Plugging the scaling relations, (2.18) and (2.21), into the loop Eq. (2.30) and taking the limit, $\varepsilon \to 0$, one obtains

$$\partial^2_{Z_{\text{cdt}}} \left[G_s W^{(3)}_{\text{GCDT}}(Z_{\text{cdt}})^2 - \partial_{Z_{\text{cdt}}} \left(\lim_{\varepsilon \to 0} \frac{V(z)}{\varepsilon^4} \right) W^{(3)}_{\text{GCDT}}(Z_{\text{cdt}}) \right] = -\frac{2}{9}, \tag{2.31}$$

where

$$\partial_{Z_{\text{cdt}}} \left(\lim_{\varepsilon \to 0} \frac{V(z)}{\varepsilon^4} \right) = \Lambda_{\text{cdt}} + \frac{1}{9} Z^3_{\text{cdt}} \equiv \partial_{Z_{\text{cdt}}} \mathscr{V}(Z_{\text{cdt}}); \tag{2.32}$$

$$w(g,z) = \varepsilon^{-1} W^{(3)}_{\text{GCDT}}(Z_{\text{cdt}}) + \mathscr{O}(\varepsilon^0). \tag{2.33}$$

This implies the existence of field theory defined by the potential $\mathscr{V}(Z_{\text{cdt}})$. Thus, the loop equation in the continuum limit becomes

$$G_s W^{(3)}_{\text{GCDT}}(Z_{\text{cdt}})^2 = \partial_{Z_{\text{cdt}}} \mathscr{V}(Z_{\text{cdt}}) W^{(3)}_{\text{GCDT}}(Z_{\text{cdt}}) - Q(Z_{\text{cdt}}), \tag{2.34}$$

[1] This value is considered to be the gravity-dressed edge singularity of the dimer model [2]. In the dimer model, $\sigma = 1/6$.

where $Q(Z_{\rm cdt})$ is a polynomial of degree 2. The solution of the loop equation (2.34) is the continuous disk function of the generalized CDT coupled to dimers. When prohibiting spatial topology change, i.e., $G_s \to 0$, one can recover the continuous disk function of CDT coupled to dimers (2.24). From this matrix model in the continuum with the potential $\mathscr{V}$, the string field theory of the generalized CDT can be constructed [7] (see Sect. 2.2.1.1 for the string field theory of the generalized CDT).

2.1.3 Discussion

The multicritical model of CDT discussed in this section is the first analytic example of CDT coupled to matter.[2] Overviewing some unsolved problems in DT and CDT, we will mention the status of our model.

Only a few riddles are left in 2d Euclidean quantum gravity coupled to matter. One of them is the behavior of the Hausdorff dimension d_h as a function of the central charge c of the conformal theory coupled to gravity. A formula was derived by Watabiki some years ago [8]

$$d_h = 2\frac{\sqrt{49-c}+\sqrt{25-c}}{\sqrt{25-c}+\sqrt{1-c}}. \tag{2.35}$$

Most likely this formula is correct for $c \leq 0$. For $c = 0$ agrees with what is known to be the correct answer [10–12]. For $c = -2$ there are very reliable computer simulations which show agreement with the formula [13, 14]. Finally for $c \to -\infty$ it gives 2, something one would indeed expect from semiclassical Liouville theory. However, for $0 < c \leq 1$ the numerical agreement is less conclusive [15, 16], and the possibility that $d_h = 4$ in this range was pointed out, and the idea has recently been resurrected [17]. For $c > 1$ the Liouville formulas become complex and expressions like (2.35) are not valid, but it is believed that there is a universal phase where the world sheet degenerates to branched polymers (BP).

Surprisingly we have a somewhat similar situation in CDT: from numerical simulations d_h seems unchanged (and equal 2) when matter with central charge $0 \leq c \leq 1$ is coupled to the CDT ensemble [18–20] and recently it was shown that there might be a kind of universal phase for $c > 1$ [21]. However, to the extent we can really view the hard dimer models as corresponding to conformal field theories, it seems that for $c < 0$ the matter systems can change fractal structure of the CDT ensemble. Qualitatively the changes are like in the full Euclidean models, d_h decreases with decreasing c. In the $c = 0$ case the CDT model can be understood as an effective Euclidean model, where baby universes have been integrated out. Whether such an interpretation is possible also when matter is coupled to CDT is presently unknown,

2 While completing the article [4] we were informed by Zohren that he and Atkin [9] have obtained results which are identical to some of our results. We thank Stefan for informing us of these results prior to publication.

but since the multicritical model captures the critical behavior of both CDT and ordinary 2d Euclidean quantum gravity coupled to certain matter systems, depending on how we scale g_s, we have a chance to answer this question in the context of analytic models like the one discussed here.

2.2 Extended Interactions in CDT

In this section, we quest for possibilities to extend the generalized CDT without changing scaling dimensions of space and time in 2 dimensions [22]: we extend the generalized CDT applying the method in the non-critical string field theory (SFT) techniques in [23, 24]; we solve the Schwinger-Dyson equation (SDE) for the disk amplitude by a perturbation of the string coupling constant. We also show dual matrix models in the continuum limit. The aim of our work [22] is to propose CDT coupled to the Ising model.

2.2.1 *Generalized CDT*

2.2.1.1 From String Field Theory

We review the non-critical string field theory (SFT) of the generalized CDT formulated in [25]. This model reproduces the disk amplitude derived in the continuum limit of CDT in the case that the string coupling constant is zero. The model requires closed strings with length L are created and annihilated from the vacuum, $|0\rangle$ ($\langle 0|$) by operators, $\psi^\dagger(L)$ and $\psi(L)$, respectively:

$$\langle 0|\psi^\dagger(L) = \psi(L)|0\rangle = 0. \tag{2.36}$$

These creation and annihilation operators obey the following commutation relation:

$$[\psi(L), \psi^\dagger(L')] = \delta(L - L'), \tag{2.37}$$

and others are zero. The string world-sheet can be seen as the Universe in 2 dimensions. Corresponding Hamiltonian can be written as follows:

$$H_0 = \int_0^\infty dL \psi^\dagger(L)\mathscr{H}_0(L, \Lambda_{\text{cdt}})\psi(L)$$
$$+ G_s \int_0^\infty dL_1 \int_0^\infty dL_2 \psi^\dagger(L_1)\psi^\dagger(L_2)\psi(L_1 + L_2)(L_1 + L_2)$$

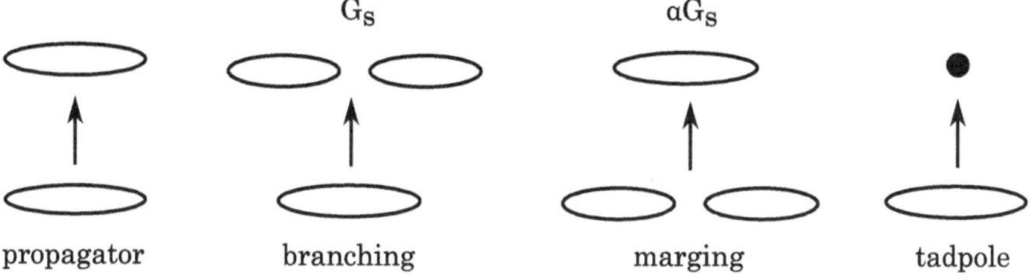

Fig. 2.3 Hamiltonian of generalized CDT

$$+\alpha G_s \int_0^\infty dL_1 \int_0^\infty dL_2 \psi^\dagger(L_1+L_2)\psi(L_2)\psi(L_1)L_2L_1 - \int_0^\infty dL\delta(L)\psi(L), \tag{2.38}$$

where

$$\mathscr{H}_0(L, \Lambda_{\mathrm{cdt}}) = -L\partial_L^2 + \Lambda_{\mathrm{cdt}} L. \tag{2.39}$$

G_s and Λ_{cdt} are the renormalized string coupling constant and the bulk cosmological constant, respectively (see Fig. 2.3); ∂_L is the derivative with respect to L. The parameter α in (2.38) was introduced to count the number of genus. In the following discussion we will take $\alpha = 0$, which suppresses the creation of genus. The Hamiltonian above can be determined under the following scaling dimensions:

$$[S] = \varepsilon, \quad [\psi^\dagger(L)] = \varepsilon^0, \quad [\psi(L)] = \varepsilon^{-1}, \quad [G_s] = \varepsilon^{-3}, \tag{2.40}$$

where ε is the scaling dimension of L, and $[S]$ is the scaling dimension of time. A crucial difference between the Hamiltonian of the non-critical SFT constructed by Ishibashi and Kawai [23] and that of generalized CDT is the existence of propagator term, $\int dL\psi^\dagger(L)\mathscr{H}_0\psi(L)$: it exists in the generalized CDT, but in IK's theory there is no such a term. This difference comes from the fact that both theories have quite different definition of time.

The authors in [25] derived the Schwiner-Dyson equation (SDE) for the disk amplitude Laplace-transformed, $W_{\mathrm{GCDT}}(Z_{\mathrm{cdt}}) = \int_0^\infty dLe^{-LZ_{\mathrm{cdt}}}\langle 0|e^{-SH_0}\ \psi^\dagger(L)|0\rangle|_{S\to\infty}$, in the generalized CDT as[3]:

$$\partial_{Z_{\mathrm{cdt}}}\big[(\Lambda_{\mathrm{cdt}} - Z_{\mathrm{cdt}}^2)W_{\mathrm{GCDT}}(Z) + G_s W_{\mathrm{GCDT}}(Z_{\mathrm{cdt}})^2\big]+1 = 0. \tag{2.41}$$

The solution was derived by a perturbative expansion w.r.t. the string coupling constant as well [25]:

[3] The authors derived the more general result with arbitrary α, but here we restricted our situation to that with $\alpha = 0$.

$$W_{\mathrm{GCDT}}(Z_{\mathrm{cdt}}) = \frac{1}{Z_{\mathrm{cdt}} + \sqrt{\Lambda_{\mathrm{cdt}}}} - G_s \frac{Z_{\mathrm{cdt}} + 3\sqrt{\Lambda_{\mathrm{cdt}}}}{4\Lambda_{\mathrm{cdt}}(Z + \sqrt{\Lambda_{\mathrm{cdt}}})^3} + \mathcal{O}(G_s^2). \quad (2.42)$$

The first term is equivalent to the CDT solution [3]. In this formalism, the contributions from baby universes are weighted by the string coupling constant G_s.

2.2.1.2 From Matrix Model

A new scaling limit of the hermitian $N \times N$ matrix model was introduced [26]. We start with the matrix integral,

$$\int d\phi \, \exp\left[-\frac{N}{g_s}\mathrm{tr}\left(\frac{1}{2}\phi^2 - g\phi - \frac{g}{3}\phi^3\right)\right] = \int d\phi \, e^{-\frac{N}{g_s}\mathrm{tr}V(\phi)}, \quad (2.43)$$

where ϕ, g and g_s are the $N \times N$ hermitian matrix, 'tHooft coupling constant and string coupling constant, respectively. One can expand the coupling constants around the critical point found in [26], using the lattice spacing ε:

$$g_s = \frac{1}{2}\varepsilon^3 G_s, \quad \phi = \hat{I} - \varepsilon\Phi + \mathcal{O}(\varepsilon^2), \quad g = \frac{1}{2}\left(1 - \frac{1}{2}\varepsilon^2 \Lambda_{\mathrm{cdt}} + \mathcal{O}(\varepsilon^4)\right), \quad (2.44)$$

where $\hat{I}$ is the unit $N \times N$ matrix; G_s, Φ and Λ_{cdt} are corresponding renormalized values. Substituting the fine-tuned values above into the potential $\frac{N}{g_s}V(\phi)$, one finds

$$\frac{N}{g_s}\mathrm{tr}V(\phi) = \frac{N}{G_s}\mathrm{tr}\left(\frac{1}{3}\Phi^3 - \Lambda_{\mathrm{cdt}}\Phi\right) + (\text{terms independent of } \Phi) + \mathcal{O}(\varepsilon). \quad (2.45)$$

Since the potential term scales at the new critical point as well as the "singular term" with fractional power, a field theory description can be anticipated. One can define it as the matrix model in the continuum limit [26, 27]. It has the following partition function:

$$Z = \int d\Phi \, \exp\left[-\frac{N}{G_s}\mathrm{tr}\left(\frac{1}{3}\Phi^3 - \Lambda_{\mathrm{cdt}}\Phi\right)\right]. \quad (2.46)$$

In the large-N limit, the saddle-point equation coincides with the SDE of the generalized CDT (2.41).[4]

[4] In [27], the authors derived the general saddle-point equation beyond the large-N limit. The general saddle-point equation indeed coincides with the SDE with arbitrary α by the treatment, $\alpha = 1/N^2$.

2.2.2 *Generalized CDT with Extended Interactions*

2.2.2.1 From String Field Theory

Applying the method in [24], we will construct the non-critical SFT Hamiltonian of the generalized CDT with extended interactions. The propagator term in (2.38), $\int dL\psi^{\dagger}(L)\mathscr{H}_0(L,\Lambda_{\mathrm{cdt}})\psi(L)$, induces causal geometries. Letting this propagator unchanged, one should take the scaling dimension of space and time as

$$[L]=\varepsilon,\quad [S]=\varepsilon. \tag{2.47}$$

We extend the non-critical SFT of the generalized CDT such that the scaling above is unchanged. Since we think that the causality is an identity of CDT, this sort of extension is meaningful to get some deep understanding of what CDT is. First, we consider strings with different charges: the $(+)$-type and $(-)$-type. The creation and annihilation operators for the $(+)$-type string, $\Psi_+^{\dagger}(L)$ and $\Psi_+(L)$, and for the $(-)$-type string, $\Psi_-^{\dagger}(L)$ and $\Psi_-(L)$, are defined as the following vacuum conditions, respectively:

$$\langle 0|\psi_+^{\dagger}(L)\ =\psi_+(L)|0\rangle=\langle 0|\psi_-^{\dagger}(L)\ =\psi_-(L)|0\rangle=0. \tag{2.48}$$

We assume these operators obey the following commutation relations:

$$[\psi_+(L),\psi_+^{\dagger}(L')]=[\psi_-(L),\psi_-^{\dagger}(L')]=\delta(L-L'), \tag{2.49}$$

and the others are zero. Additionally, we assume the same scaling dimensions with those of the generalized CDT:

$$[\psi_{\pm}^{\dagger}(L)]=\varepsilon^{0},\quad [\psi_{\pm}(L)]=\varepsilon^{-1},\quad [G_s]=\varepsilon^{-3}, \tag{2.50}$$

where G_s is the string coupling constant. Under the conditions above, one can extend the Hamiltonian of the generalized CDT applying the interaction for spin clusters introduced by Ishibashi and Kawai [24]. Here we call such an interaction *IK-type interaction*. It is based on the so-called *peeling procedure* in a discrete random surface. For example, considering a randomly triangulated surface coupled to Ising spins with one boundary and then assuming that triangles attached to the boundary have homogeneous spins (all spins are up-type or down-type), one peels triangles along with the boundary as if one peels an apple. If one continues to peel off triangles over the boundary triangles and one encounters the triangle carrying a different type of spin, then one surrounds them by the triangles carrying the same spin as the boundary triangles. In short, the randomly triangulated surface is separated by domain walls. The SDE in their approach coincides with the loop equation for the chain-type two-matrix model describing random geometries coupled to Ising spins. We emphasize here that the above closed strings are not seen as the spin boundary as in the case

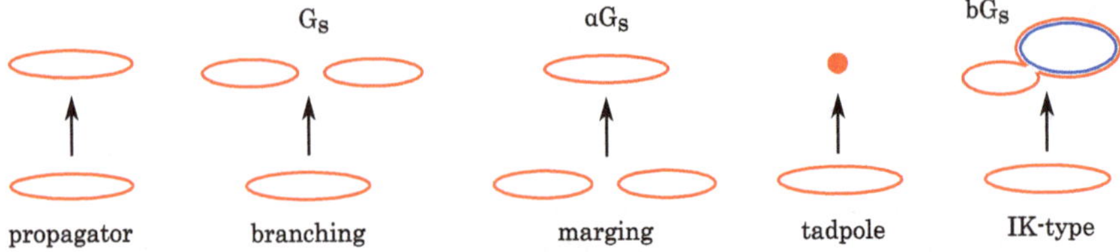

Fig. 2.4 Hamiltonian of generalized CDT with extended interactions

of IK but seen as the equal-time hypersurfaces with different charges. Applying the IK-type interaction, one can write the extended Hamiltonian of the generalized CDT:

$$\begin{aligned} H_m = &\int_0^\infty dL \psi_+^\dagger(L) \mathscr{H}_0(L, \Lambda_{\mathrm{cdt}}) \psi_+(L) \\ &+ G_s \int_0^\infty dL_1 \int_0^\infty dL_2 \psi_+^\dagger(L_1) \psi_+^\dagger(L_2) \psi_+(L_1 + L_2)(L_1 + L_2) \\ &+ bG_s \int_0^\infty dL_1 \int_0^\infty dL_2 \psi_+^\dagger(L_1 + L_2) \psi_-^\dagger(L_2) \psi_+(L_1) L_1 \\ &+ \alpha G_s \int_0^\infty dL_1 \int_0^\infty dL_2 \psi_+^\dagger(L_1 + L_2) \psi_+(L_2) \psi_+(L_1) L_2 L_1 \\ &- \int_0^\infty dL \delta(L) \psi_+(L) + \left[\psi_+(\psi_+^\dagger) \leftrightarrow \psi_-(\psi_-^\dagger) \right], \end{aligned} \tag{2.51}$$

where α and b are dimension-less constants (see Fig. 2.4).[5]

For simplicity, we restrict topology to the disk. This can be realized by the following Hamiltonian:

$$H_m^D \equiv \lim_{\alpha \to 0} H_m. \tag{2.52}$$

We will then derive the SDE in our model. The SDE corresponds to the Wheeler-DeWitt equation for the wave function of the Universe. We define the partition function and disk amplitudes:

[5] In fact, it is possible to include the interactions, $\int dL \psi_-^\dagger(L) \mathscr{H}_0(L, \Lambda_{\mathrm{cdt}}) \psi_+(L)$ and its spin-flipped term. However, because of the $\mathbb{Z}_2$-symmetry as to the spin reflection, such terms merely cause a constant shift of the string coupling constant, so that we have not included these terms in the Hamiltonian.

$$Z = \lim_{S\to\infty} \langle 0|e^{-SH_m^D}|0\rangle \equiv 1, \tag{2.53}$$

and

$$W_\pm(L) \equiv \lim_{S\to\infty} \langle 0|e^{-SH_m^D}\psi_\pm^\dagger(L)|0\rangle. \tag{2.54}$$

The SDE for $W_\pm(L)$ is

$$\lim_{S\to\infty} \frac{\partial}{\partial S}\langle 0|e^{-SH_m^D}\psi_\pm^\dagger(L)|0\rangle = 0. \tag{2.55}$$

Using the equation, $H_m^D|0\rangle = 0$, and the commutation relations (2.49), one can rewrite the SDE as follows:

$$\begin{aligned} 0 = &- L\partial_L^2 W_\pm(L) + \Lambda L W_\pm(L) - \delta(L) \\ &+ G_s L \int_0^\infty dL_1 \lim_{S\to\infty} \langle 0|e^{-SH_m^D}\psi_\pm^\dagger(L_1)\psi_\pm^\dagger(L-L_1)|0\rangle \\ &+ bG_s L \int_0^\infty dL_1 \lim_{S\to\infty} \langle 0|e^{-SH_m^D}\psi_\pm^\dagger(L+L_1)\psi_\mp^\dagger(L+L_1)|0\rangle. \end{aligned} \tag{2.56}$$

Here we introduce the factorization theorem:

$$\begin{aligned} \lim_{S\to\infty} \langle 0|e^{-SH_m^D}\psi_\pm^\dagger(L_1)\psi_\pm^\dagger(L_2)|0\rangle = &\lim_{S\to\infty} \langle 0|e^{-SH_m^D}\psi_\pm^\dagger(L_1)|0\rangle \lim_{S\to\infty} \langle 0|e^{-SH_m^D} \\ &\psi_\pm^\dagger(L_2)|0\rangle. \end{aligned} \tag{2.57}$$

Applying the above factorization theorem, the SDE (2.57) becomes

$$\begin{aligned} 0 = &-L\partial_L^2 W_\pm(L) + \Lambda_{\mathrm{cdt}} L W_\pm(L) - \delta(L) + G_s L \int_0^\infty dL_1 W_\pm(L_1) W_\pm(L-L_1) \\ &+ bG_s L \int_0^\infty dL_1 W_\pm(L+L_1) W_\mp(L_1). \end{aligned} \tag{2.58}$$

In fact, our system has $\mathbb{Z}_2$-symmetry w.r.t. the spin-reflection. Thus, we focus on a $\mathbb{Z}_2$-symmetric solution of the SDE:

$$W_+(L) = W_-(L) \equiv W_{\mathrm{IK}}(L). \tag{2.59}$$

Implementing the Laplace transformation, $\mathscr{L}[W_{\mathrm{IK}}(L)] \equiv \int_0^\infty dL e^{-LZ_{\mathrm{cdt}}} W_{\mathrm{IK}}(L) \equiv W_{\mathrm{IK}}(Z_{\mathrm{cdt}})$, (2.58) becomes

$$0 = \partial_{Z_{\mathrm{cdt}}}\left[(Z_{\mathrm{cdt}}^2 - \Lambda_{\mathrm{cdt}})W_{\mathrm{IK}}(Z_{\mathrm{cdt}}) - G_s W_{\mathrm{IK}}(Z_{\mathrm{cdt}})^2\right] - 1 + bG_s\mathscr{L}\left[L\int dL_1 W_{\mathrm{IK}}(L+L_1)W_{\mathrm{IK}}(L_1)\right]. \tag{2.60}$$

One notices that the last term diverges in $Z \to \infty$. To regularize this divergence, we symmetrize the term w.r.t. the reflection, $Z_{\mathrm{cdt}} \leftrightarrow -Z_{\mathrm{cdt}}$ [24, 28–30]:

$$\int_0^\infty dL \int_0^\infty dL_1 e^{-Z_{\mathrm{cdt}}(L+L_1)} W_{\mathrm{IK}}(L+L_1) e^{+Z_{\mathrm{cdt}}L_1} W_{\mathrm{IK}}(L_1) + (Z_{\mathrm{cdt}} \leftrightarrow -Z_{\mathrm{cdt}}) = W_{\mathrm{IK}}(Z_{\mathrm{cdt}})W_{\mathrm{IK}}(-Z_{\mathrm{cdt}}). \tag{2.61}$$

Subtracting the SDE with the reflection ($Z \to -Z$) from the original SDE (2.60), one obtains the finite SDE:

$$\partial_{Z_{\mathrm{cdt}}}\Big[(Z_{\mathrm{cdt}}^2 - \Lambda_{\mathrm{cdt}})\left(W_{\mathrm{IK}}(Z_{\mathrm{cdt}}) + W_{\mathrm{IK}}(-Z_{\mathrm{cdt}})\right) - G_s\left(W_{\mathrm{IK}}(Z_{\mathrm{cdt}})^2 + W_{\mathrm{IK}}(-Z_{\mathrm{cdt}})^2 + bW_{\mathrm{IK}}(Z_{\mathrm{cdt}})W_{\mathrm{IK}}(-Z_{\mathrm{cdt}})\right)\Big] = 0. \tag{2.62}$$

Integrating the SDE above over Z_{cdt} yields

$$(Z_{\mathrm{cdt}}^2 - \Lambda_{\mathrm{cdt}})\left(W_{\mathrm{IK}}(Z_{\mathrm{cdt}}) + W_{\mathrm{IK}}(-Z_{\mathrm{cdt}})\right) - G_s\left(W_{\mathrm{IK}}(Z_{\mathrm{cdt}})^2 + W_{\mathrm{IK}}(-Z_{\mathrm{cdt}})^2 + bW_{\mathrm{IK}}(Z_{\mathrm{cdt}})W_{\mathrm{IK}}(-Z_{\mathrm{cdt}})\right) = c. \tag{2.63}$$

where c is a constant. We calculate a perturbative solution for the SDE above around the weak coupling region, $G_s < 1$, expanding the loop amplitude and c like:

$$W_{\mathrm{IK}}(Z_{\mathrm{cdt}}) = \sum_{n=0}^\infty G_s^n W_n(Z_{\mathrm{cdt}}), \quad c = \sum_{n=0}^\infty G_s^n c_n. \tag{2.64}$$

As for $W_0(Z_{\mathrm{cdt}})$, one finds

$$W_0(Z_{\mathrm{cdt}}) = \frac{1}{Z_{\mathrm{cdt}} + \sqrt{\Lambda_{\mathrm{cdt}}}}, \tag{2.65}$$

where we have chosen an overall constant for $W_0(Z)$ so as to coincide with that of the plain CDT. Assuming that the disk amplitude behaves as $1/Z_{\mathrm{cdt}}$ in the large value of $|Z_{\rm cdt}|$, we can determine that $c_1 = -(b+1)/2\Lambda_{\rm cdt}$. Furthermore, we can extract $W_1(Z_{\rm cdt})$ considering that $W_1(Z_{\rm cdt})$ is analytic in the region, $|Z_{\rm cdt}| > 0$. Thus, the perturbative solution is

$$W_{\rm IK}(Z_{\rm cdt}) = \frac{1}{Z_{\rm cdt}+\sqrt{\Lambda_{\rm cdt}}} - G_s\frac{1}{4\Lambda_{\rm cdt}}\left[\frac{Z_{\rm cdt}+3\sqrt{\Lambda_{\rm cdt}}}{(Z_{\rm cdt}+\sqrt{\Lambda_{\rm cdt}})^3} + \frac{b}{(Z_{\rm cdt}+\sqrt{\Lambda_{\rm cdt}})^2}\right] + \mathcal{O}(G_s^2). \tag{2.66}$$

The solution with $b = 0$ is equivalent to that of the plain generalized CDT (2.42).

2.2.2.2 From Matrix Model

We start with the following matrix integral:

$$\int d\phi_+ d\phi_- e^{-\frac{N}{g_s}\mathrm{tr}V(\phi_+,\phi_-)}, \tag{2.67}$$

where

$$V(\phi_+,\phi_-) = \frac{1}{2}(\phi_+^2+\phi_-^2) - g(\phi_+ + \phi_-) - \frac{g}{3}(\phi_+^3+\phi_-^3) + x\phi_+\phi_-. \tag{2.68}$$

In the integral above, $\phi_\pm$, g, g_s and x are $N \times N$ hermitian matrices, the 'tHooft coupling constant, string coupling constant and coupling constant characterizing the interaction, respectively. We then expand the fields and coupling constants w.r.t. the lattice spacing ε as follows:

$$\phi_+ = \hat{I} - \varepsilon(A+B) + \mathcal{O}(\varepsilon^2), \quad \phi_- = \hat{I} - \varepsilon(A-B) + \mathcal{O}(\varepsilon^2), \tag{2.69}$$

and

$$g_s = \varepsilon^3 G_s, \quad g = \frac{1}{2}\left(1 - \frac{1}{2}\varepsilon^2(\Lambda_{\rm cdt} - 2X) + \mathcal{O}(\varepsilon^4)\right), \quad x = X\varepsilon^2, \tag{2.70}$$

where A and B are $N \times N$ hermitian matrices; $\hat{I}$ is the unit matrix; G_s, Φ, Λ and X are corresponding renormalized values. This construction implies that the cut-length shrinks to zero ($g_s \to 0$), and the strength of the interaction falls off ($x \to 0$). The causality induces the scaling, $g_s \to 0$ and taking the limit, $x \to 0$, the model can be seen as the weakly interacting model. Substituting the fine-tuned values, one can write the partition function of the matrix model in the continuum limit:

$$Z = \int dAdB \exp\left[-\frac{N}{G_s}\mathrm{tr}\left(\frac{1}{3}A^3 + AB^2 - \Lambda_{\rm cdt}A\right)\right]. \tag{2.71}$$

An interesting thing is that in the matrix model having this type of potential, the Gaussian integral over B can be performed by introducing the eigenvalues λ_i's for the matrix A [31]:

$$Z \propto \int \prod_i d\lambda_i \prod_{i<j} (\lambda_j - \lambda_i)^2 \prod_{i,j} (\lambda_i + \lambda_j)^{-1/2} e^{-\frac{N}{G_s} V}, \tag{2.72}$$

where

$$V = \sum_{i=1}^{N} V(\lambda_i) = \sum_{i=1}^{N} \left(\frac{1}{3} \lambda_i^3 - \Lambda_{\mathrm{cdt}} \lambda_i \right). \tag{2.73}$$

In the large-N limit, the saddle point equation becomes

$$\frac{2}{N} \sum_{j \neq i} \frac{1}{\lambda_i - \lambda_j} = \frac{1}{N} \sum_j \frac{1}{\lambda_i + \lambda_j} + \frac{1}{G_s} V'(\lambda_i), \tag{2.74}$$

where $V'(\lambda_i) = \lambda_i^2 - \Lambda_{\mathrm{cdt}}$. Here we define the resolvent for A as $W_{\mathrm{IK}}(Z_{\mathrm{cdt}}) \equiv \frac{1}{N} \mathrm{tr}(Z_{\mathrm{cdt}} - A)^{-1}$, and the distribution of eigenvalues as $\rho(\lambda) \equiv \frac{1}{N} \sum_i \delta(\lambda - \lambda_i)$. Multiplying (2.74) by $1/(Z - \lambda_i)$ and summing over i, one obtains the loop equation in the large-N limit:

$$\begin{aligned} &(Z_{\mathrm{cdt}}^2 - \Lambda_{\mathrm{cdt}}) \left(W_{\mathrm{IK}}(Z_{\mathrm{cdt}}) + W_{\mathrm{IK}}(-Z_{\mathrm{cdt}}) \right) \\ &\quad - G_s \left(W_{\Lambda_{\mathrm{cdt}}}(Z_{\mathrm{cdt}})^2 + W_{\mathrm{IK}}(Z_{\mathrm{cdt}}) W_{\mathrm{IK}}(-Z_{\mathrm{cdt}}) + W_{\mathrm{IK}}(-Z_{\mathrm{cdt}})^2 \right) = 2 \int d\lambda \rho(\lambda) \lambda. \end{aligned} \tag{2.75}$$

Remember the SDE derived in the non-critical SFT approach (2.66). One can find a great similarity between the two. Namely, when setting $b = 1$ in the SDE, the two equations are exactly same. Thus, this matrix model in the continuum limit can reproduce the generalized CDT with extended interactions in $b = 1$.

One can extend the matrix model in the continuum limit above to the general $O(n)$ vector model [31] such that:

$$Z = \int dA dB_1 \ldots dB_n e^{-\frac{N}{G_s} \mathrm{tr} U(A, B_1, \ldots, B_n)}, \tag{2.76}$$

where

$$U(A, B_1, \ldots, B_n) = A(B_1^2 + \cdots + B_n^2) + \frac{1}{3} A^3 - \Lambda_{\mathrm{cdt}} A; \tag{2.77}$$

A, B_1, ..., B_n are $N \times N$ hermitian matrices. Notice that the previous matrix model in the continuum limit is $O(1)$ vector model. Integrating out all B_i's, the loop equation in the large-N limit is

$$(Z_{\mathrm{cdt}}^2 - \Lambda_{\mathrm{cdt}})\,(W_{\mathrm{IK}}(Z_{\mathrm{cdt}}) + W_{\mathrm{IK}}(-Z_{\mathrm{cdt}}))$$
$$- G_s \left(W_{\Lambda_{\mathrm{cdt}}}(Z_{\mathrm{cdt}})^2 + n W_{\mathrm{IK}}(Z_{\mathrm{cdt}}) W_{\mathrm{IK}}(-Z_{\mathrm{cdt}}) + W_{\mathrm{IK}}(-Z_{\mathrm{cdt}})^2 \right) = 2 \int d\lambda \rho(\lambda)\lambda. \tag{2.78}$$

The loop equation of the $O(n)$ vector model coincides with the SDE labeled by the free parameter b (2.63) only if one identifies n with b.

2.2.3 Discussion

We have shown the equivalence between the two different field theories at the level of differential equations, the Schwinger-Dyson equation in the non-critical SFT and the loop equation of the matrix model in the continuum limit. In the following, we will examine the extended models from different point of view.

To begin with, we discuss our model in terms of the SFT approach. Although we have used the IK-type interaction to construct the extended SFT of the generalized CDT, we do not understand if our model is on the critical point of the Ising model characterized by the Curie temperature. In the following, we try to explain two complications around this issue. First, at the critical point of Ising spins the spin configuration must be random: spins are supposed to fluctuate all length scales between the lattice spacing and the correlation length. Contrary to that, in our model the homogeneous spin (charge) configurations survive as the propagators. Second, the definition of time associated with our Hamiltonian (2.51) is different from the would-be generalized CDT coupled to Ising spins. Namely, we consider the closed strings in our model as not spin-cluster boundaries but spatial boundaries, so that we pursue the time flow of spatial boundaries. Thus, our time is nothing but the proper time. This proper time is crucially different from the time defined via the spin-cluster boundary [32, 33]. Considering our time as the one defined via the spin-cluster boundary is equivalent to treating our model as the generalized CDT coupled to Ising spins; the scaling dimension of time may be different from the lattice spacing ε according to [32]. This contradicts our first setup (2.47). The free parameter b, in one way or another, might be the key to know what our model is.

In addition, it is possible to extend our non-critical SFT to the multi-"colored" system:

$$H_m^{(n)} = \sum_{i=1}^{n} \int_0^\infty dL \psi_i^\dagger(L) \mathscr{H}_0(L, \Lambda_{\mathrm{cdt}}) \psi_i(L)$$
$$+ G_s \sum_{i=1}^{n} \int_0^\infty dL_1 \int_0^\infty dL_2 \psi_i^\dagger(L_1) \psi_i^\dagger(L_2) \psi_i(L_1 + L_2)(L_1 + L_2)$$

$$
\begin{aligned}
&+G_s \sum_{i=1}^{n} \sum_{j \neq i}^{n} b_{ij} \int_0^\infty dL_1 \int_0^\infty dL_2 \psi_i^\dagger(L_1+L_2) \psi_j^\dagger(L_2) \psi_i(L_1) L_1 \\
&+\alpha G_s \sum_{i=1}^{n} \int_0^\infty dL_1 \int_0^\infty dL_2 \psi_i^\dagger(L_1+L_2) \psi_i(L_2) \psi_i(L_1) L_2 L_1 \\
&-\sum_{i=1}^{n} \int_0^\infty dL \delta(L) \psi_i(L). \qquad (2.79)
\end{aligned}
$$

One can derive the free parameter b in our model from the multi-"colored" system under the treatment, $W_1(L) = \cdots = W_n(L) \equiv W_\Lambda(L)$; $b_{ij} = 0$ for $j = i$; $b_{ij} = 1$ for $j \neq i$.

Next, we closely look at our matrix model. We consider the direct product of the two copies of the potential. Each of them yields the plain generalized CDT. Introducing the linear combinations of matrices such that $\Phi_+ = A + B$ and $\Phi_- = A - B$, one finds

$$
\frac{1}{\tilde{G}_s}\left(\frac{1}{3}\Phi_+^3 - \Lambda_{\rm cdt}\Phi_+ + \frac{1}{3}\Phi_-^3 - \Lambda_{\rm cdt}\Phi_-\right) = \frac{1}{G_s}\left(\frac{1}{3}A^3 + AB^2 - \Lambda_{\rm cdt}A\right), \qquad (2.80)
$$

where $\tilde{G}_s = 2G_s$. This is nothing but the potential of our $O(1)$ vector model in the continuum limit. Diagonalizing the matrix A as $A = \mathrm{diag}(\lambda_1, \ldots, \lambda_N)$ and integrating out the matrix B, one gets the effective theory for the eigenvalues of A with the potential,

$$
\underbrace{\left[\frac{1}{G_s}\sum_i\left(\frac{1}{3}\lambda_i^3 - \Lambda_{\rm cdt}\lambda_i\right) - \frac{1}{N}\log\prod_{i<j}(\lambda_j - \lambda_i)^2\right]}_{\text{terms appeared in the plain generalized CDT}}
+ \text{terms induced by the integration over } B. \qquad (2.81)
$$

An important point here is that our model is slightly different from the plain generalized CDT matrix model because integrating out the matrix B the extra correction is added to terms appeared in the plain generalized CDT. From the matrix A's point of view, the matrix B can be seen like some external field. The strength of such an external field can be bigger by inserting the integrated-out matrices. It leads to the $O(n)$ vector model in the continuum limit.

References

1. Kazakov, V. A. (1986). Ising model on a dynamical planar random lattice: Exact solution. *Physics Letter A*, *119*, 140.
2. Staudacher, M. (1990). The Yang-lee edge singularity on a dynamical planar random surface. *Nuclear Physics B*, *336*, 349.

3. Ambjorn, J., & Loll, R. (1998). Nonperturbative Lorentzian quantum gravity, causality and topology change. *Nuclear Physics B, 536*, 407. [hep-th/9805108].
4. Ambjorn, J., Glaser, L., Gorlich, A., & Sato, Y. (2012). New multicritical matrix models and multicritical 2d CDT. *Physics Letter B, 712*, 109. arXiv:1202.4435 [hep-th].
5. Moore, G. W., Seiberg, N., & Staudacher, M. (1991). *Nuclear Physics B, 362*, 665.
6. Glaser, L. (2012). Coupling dimers to CDT. arXiv:1210.4063 [hep-th].
7. Atkin, M. R., & Zohren, S. (2012). On the quantum geometry of multi-critical CDT. *JHEP, 1211*, 037. arXiv:1203.5034 [hep-th].
8. Watabiki, Y. (1993). Analytic study of fractal structure of quantized surface in two-dimensional quantum gravity. *Progress of Theoretical Physics Supplement, 114*, 1.
9. Atkin, M. R., & Zohren, S. (2012). An analytical analysis of CDT coupled to dimer-like matter. *Physics Letter B, 712*, 445. arXiv:1202.4322 [hep-th].
10. Kawai, H., Kawamoto, N., Mogami, T., & Watabiki, Y. (1993). Transfer matrix formalism for two-dimensional quantum gravity and fractal structures of space-time. *Physics Letter B, 306*, 19. [hep-th/9302133].
11. Ambjorn, J., & Watabiki, Y. (1995). Scaling in quantum gravity. *Nuclear Physics B, 445*, 129. [hep-th/9501049].
12. Aoki, H., Kawai, H., Nishimura, J., & Tsuchiya, A. (1996). Operator product expansion in two-dimensional quantum gravity. *Nuclear Physics B, 474*, 512. [hep-th/9511117].
13. Ambjorn, J., Anagnostopoulos, K. N., Ichihara, T., Jensen, L., Kawamoto, N., Watabiki, Y., et al. (1997). Quantum geometry of topological gravity. *Physics Letter B, 397*, 177. [hep-lat/9611032].
14. Ambjorn, J., Anagnostopoulos, K., Ichihara, T., Jensen, L., Kawamoto, N., Watabiki, Y., et al. (1998). The quantum space-time of $c = -2$ gravity. *Nuclear Physics B, 511*, 673. [hep-lat/9706009].
15. Ambjorn, J., & Anagnostopoulos, K. N. (1997). Quantum geometry of 2-D gravity coupled to unitary matter. *Nuclear Physics B, 497*, 445. [hep-lat/9701006].
16. Ambjorn, J., Anagnostopoulos, K. N., Magnea, U., & Thorleifsson, G. (1996). Geometrical interpretation of the KPZ exponents. *Physics Letter B, 388*, 713. [hep-lat/9606012].
17. Duplantier, B. (2011) The hausdorff dimension of two-dimensional quantum gravity. arXiv:1108.3327 [math-ph].
18. Ambjorn, J., Anagnostopoulos, K. N., Loll, R., & Pushkina, I. (2009). Shaken, but not stirred: Potts model coupled to quantum gravity. *Nuclear Physics B, 807*, 251. arXiv:0806.3506 [hep-lat].
19. Ambjorn, J., Anagnostopoulos, K. N., & Loll, R. (1999). A new perspective on matter coupling in 2-D quantum gravity. *Physics Review D, 60*, 104035. [hep-th/9904012].
20. Ambjorn, J., Anagnostopoulos, K. N., & Loll, R. (2000). Crossing the $c = 1$ barrier in 2-D Lorentzian quantum gravity. *Physics Review D, 61*, 044010. [hep-lat/9909129].
21. Ambjorn, J., Goerlich, A. T., Jurkiewicz, J., & Zhang, H.-G. (2012). Pseudo-topological transitions in 2D gravity models coupled to massless scalar fields. *Nuclear Physics B, 863*, 421. arXiv:1201.1590 [gr-qc].
22. Fuji, H., Sato, Y., & Watabiki, Y. (2011). Causal dynamical triangulation with extended interactions in $1 + 1$ dimensions. *Physics Letter B, 704*, 582. arXiv:1108.0552 [hep-th].
23. Ishibashi, N., & Kawai, H. (1993). String field theory of noncritical strings. *Physics Letter B, 314*, 190. [hep-th/9307045].
24. Ishibashi, N., & Kawai, H. (1994). String field theory of c ¡= 1 noncritical strings. *Physics Letter B, 322*, 67. [hep-th/9312047].
25. Ambjorn, J., Loll, R., Watabiki, Y., Westra, W., & Zohren, S. (2008). A string field theory based on causal dynamical triangulations. *JHEP, 0805*, 032. arXiv:0802.0719 [hep-th].
26. Ambjorn, J., Loll, R., Watabiki, Y., Westra, W., & Zohren, S. (2008). A new continuum limit of matrix models. *Physics Letter B, 670*, 224. arXiv:0810.2408 [hep-th].
27. Ambjorn, J., Loll, R., Watabiki, Y., Westra, W., & Zohren, S. (2008). A matrix model for 2D quantum gravity defined by causal dynamical triangulations. *Physics Letter B, 665*, 252. arXiv:0804.0252 [hep-th].

28. Kostov, I. K. (1989). *Nuclear Physics B*, *326*, 583.
29. Kostov, I. K. (1991). *Physics Letter B*, *266*, 42.
30. Kostov, I. K. (1992). *Nuclear Physics B*, *376*, 539.
31. Kostov, I. K. (1989). O (n) vector model on a planar random lattice: Spectrum of anomalous dimensions. *Modern Physics Letters A*, *4*, 217.
32. Ambjorn, J., Anagnostopoulos, K. N., Jurkiewicz, J., & Kristjansen, C. F. (1998). The concept of time in 2-D gravity. *JHEP*, *9804*, 016. [hep-th/9802020].
33. Ambjorn, J., Kristjansen, C., & Watabiki, Y. (1997). The two point function of $c = -2$ matter coupled to 2-D quantum gravity. *Nuclear Physics B*, *504*, 555. [hep-th/9705202].

Chapter 3
n-DBI Gravity

3.1 Black Hole Solutions in n-DBI Gravity

Considering the full n-DBI action, we have shown that any solution of Einstein gravity with a particular curvature property is a solution of n-DBI gravity [1]. Amongst them is a class of geometries isometric to Reissner-Nordström-(Anti) de Sitter black hole, which is obtained within the spherically symmetric solutions of n-DBI gravity minimally coupled to the Maxwell field. These solutions have, however, two distinct features from their Einstein gravity counterparts: (1) the cosmological constant appears as an integration constant and can be positive, negative or vanishing, making it a *variable* quantity of the theory; (2) there is a non-uniqueness of solutions with the same total mass, charge and effective cosmological constant. Such inequivalent solutions cannot be mapped to each other by a foliation preserving diffeomorphism. Physically they are distinguished by the expansion and shear of the congruence tangent to n, which define scalar invariants on each leave of the foliation.

3.1.1 Solutions with Constant $\mathscr{R}$

We start with the action of n-DBI gravity with matter [1, 2]:

$$S = -\frac{3\lambda}{4\pi G_4^2}\int dtd^3x\sqrt{h}N\left[\sqrt{1+\frac{G_4}{6\lambda}\left(R+K_{ij}K^{ij}-K^2-2N^{-1}\Delta N\right)}-q\right] - \int d^4x\mathscr{L}_{\text{matter}}, \tag{3.1}$$

where G_4 is Newton's constant and $\mathscr{L}_{\text{matter}}$ is the matter Lagrangian density. The theory includes two dimensionless constants: λ and q (as for other ingredients, see Sect. 1.3.3). Here for convenience, we choose the following convention:

Y. Sato, *Space-Time Foliation in Quantum Gravity*, Springer Theses,
DOI: 10.1007/978-4-431-54947-5_3, © Springer Japan 2014

$$R + K_{ij}K^{ij} - K^2 - 2N^{-1}\Delta N \equiv \mathscr{R}. \tag{3.2}$$

Taking variations of N, N^i and h_{ij}, one obtains the Hamiltonian constraint, momentum constrains and evolution equation, respectively:

Hamiltonian constraint:

$$\frac{1 + \frac{G_4}{6\lambda}\left(R - N^{-1}\Delta N\right)}{\sqrt{1 + \frac{G_4}{6\lambda}\mathscr{R}}} - q - \frac{G_4}{6\lambda}\Delta\left(1 + \frac{G_4}{6\lambda}\mathscr{R}\right)^{-1/2} = -\frac{4\pi G_4^2}{3\lambda\sqrt{h}}\frac{\delta\mathscr{L}_{\text{matter}}}{\delta N}, \tag{3.3}$$

Momentum constraints:

$$\nabla^j\left(\frac{K_{ij} - h_{ij}K}{\sqrt{1 + \frac{G_4}{6\lambda}\mathscr{R}}}\right) = -\frac{8\pi G_4}{\sqrt{h}}\frac{\delta\mathscr{L}_{\text{matter}}}{\delta N^i}, \tag{3.4}$$

Evolution equation:

$$\begin{aligned}
&\frac{1}{N}(\pounds_t - \pounds_N)\left(\frac{K^{ij} - h^{ij}K}{\sqrt{1 + \frac{G_4}{6\lambda}\mathscr{R}}}\right) \\
&= (\nabla^i\nabla^j - h^{ij}(\nabla_l \ln N)\nabla^l)\left(1 + \frac{G_4}{6\lambda}\mathscr{R}\right)^{-1/2} \\
&\quad + \frac{-R^{ij} + KK^{ij} - 2K^{il}K_l^j + N^{-1}\nabla^i\nabla^j N + h^{ij}(K_{mn}K^{mn} - N^{-1}\Delta N)}{\sqrt{1 + \frac{G_4}{6\lambda}\mathscr{R}}} \\
&\quad + \frac{16\pi G_4}{N\sqrt{h}}\frac{\delta\mathscr{L}_{\text{matter}}}{\delta h_{ij}}.
\end{aligned} \tag{3.5}$$

Taking $\mathscr{R}$ to be constant, we have found very interesting solutions [1]. We will explain them in the following. Dubbing $\sqrt{1 + G_4\mathscr{R}/(6\lambda)} \equiv C$, the equations of motion (3.3)–(3.5) reduce to

$$R - N^{-1}\Delta N + \frac{6\lambda}{G_4}(1 - qC) = -\frac{8\pi G_4 C}{\sqrt{h}}\frac{\delta\mathscr{L}_{\text{matter}}}{\delta N}, \tag{3.6}$$

$$\nabla^j\left(K_{ij} - h_{ij}K\right) = -\frac{8\pi G_4 C}{\sqrt{h}}\frac{\delta\mathscr{L}_{\text{matter}}}{\delta N^i}, \tag{3.7}$$

$$\frac{1}{N}(\pounds_t - \pounds_N)\left(K^{ij} - h^{ij}K\right) = h^{ij}(K_{mn}K^{mn} - N^{-1}\Delta N) \\ -R^{ij} + KK^{ij} - 2K^{il}K_l^j + N^{-1}\nabla^i\nabla^j N \\ +\frac{16\pi G_4 C}{N\sqrt{h}}\frac{\delta \mathscr{L}_{\text{matter}}}{\delta h_{ij}}. \tag{3.8}$$

The momentum constraints and the dynamical equations are equivalent to those of Einstein gravity, but with a renormalization of the matter terms by a factor of C. The Hamiltonian constraint is also equivalent to that of Einstein gravity with, besides the renormalization by C of the matter term, a cosmological constant

$$\Lambda_C = \frac{3\lambda}{G_4}(2qC - 1 - C^2). \tag{3.9}$$

We are thus led to the following theorem and corollary:

Theorem: *Any solution of Einstein gravity with a cosmological constant plus some matter Lagrangian admitting a foliation with constant $\mathscr{R}$, as defined in* (3.2), *is a solution of n-DBI gravity* (*with an appropriate renormalization of the solution parameters*).

Corollary: *Any Einstein space* (*hence solution of the Einstein equations with a cosmological constant*) *admitting a foliation such that $R - N^{-1}\Delta N$ is constant—where R and Δ are the Ricci scalar and the Laplacian of the 3-metric h_{ij} and N is the lapse in the ADM decomposition—is a solution of n-DBI gravity* (*with an appropriate renormalization of the solution parameters*).

As shown in [3], this fact can be understood naturally through the linearlized n-DBI action (1.133):

$$S^e = -\frac{1}{16\pi G_4}\int d^4x\sqrt{-g}e\left[{}^{(4)}R - 2\Lambda_C(e) + \mathscr{K}\right],$$

where

$$\Lambda_C(e) = \frac{3\lambda}{G_4}\left(\frac{2q}{e} - 1 - \frac{1}{e^2}\right). \tag{3.10}$$

Solving the equation of motion of the auxiliary field e, one finds the following solution:

$$e = \pm\left(1 + \frac{G_4}{6\lambda}\mathscr{R}\right)^{-1/2}. \tag{3.11}$$

Therefore, $e = \pm 1/C$. Picking up $+$ in the double sign, one finds that two cosmological constants, (3.9) and (3.10), coincide as advocated before. The constant $\mathscr{R}$ means the constant e, so that (1.133) becomes equivalent to the Einstein-Hilbert action with cosmological constant Λ_C and the Gibbons-Hawking-York term up to an overall coefficient. This is the reason why the theorem holds.

3.1.2 Spherically Symmetric Solutions

The most generic spherically symmetric line element reads

$$ds^2 = -g_{TT}(R,T)dT^2 + g_{RR}(R,T)dR^2 + 2g_{TR}(T,R)dTdR + g_{\theta\theta}(R,T)d\Omega_2. \quad (3.12)$$

Defining a new radial coordinate $r^2 = g_{\theta\theta}(R,T)$ and a new time coordinate $t = t(R,T)$ it is possible to transform this line element into a standard diagonal form, with only two unknown functions: $g_{tt}(t,r)$ and $g_{rr}(t,r)$. Then, the vacuum Einstein equations yield, as the only solution, the Schwarzschild black hole, and as a corollary Birkhoff's theorem, namely that spherical symmetry implies staticity. In n-DBI gravity, however, only foliation preserving diffeomorphisms are allowed. Thus, only $t = t(T)$ is allowed. As a consequence, the most general line element compatible with spherical symmetry is

$$ds^2 = -N^2(t,r)dt^2 + e^{2f(t,r)}\left(dr + e^{2g(t,r)}dt\right)^2 + r^2 d\Omega_2. \quad (3.13)$$

To include the possibility of charge, we take

$$\mathscr{L}_{\text{matter}} = -\frac{1}{16\pi}\sqrt{-g}F_{\mu\nu}F^{\mu\nu}, \quad (3.14)$$

where $\mathbf{F} = d\mathbf{A}$ is the Maxwell 2-form. In order to find solutions, we impose two assumptions: we consider the case with

1. only r dependence for the three metric functions.
2. pure electric

$$\mathbf{A} = A(r)dt \quad \Rightarrow \quad \mathscr{L}_{\text{matter}} = \frac{r^2 \sin\theta e^{-f}}{8\pi N}(A')^2. \quad (3.15)$$

It is more convenient to consider the reduced system obtained by specializing the action (3.1) to our ansatz rather than solving the full equations of motion. One obtains the effective Lagrangian density:

$$\mathscr{L}_{\text{eff}} = \lambda r^2 e^f N\left[\sqrt{1 + \frac{G_N}{6\lambda}\mathscr{R}} - q\right] + \frac{G_N^2}{6N} r^2 e^{-f}(A')^2, \quad (3.16)$$

whose equations of motion are a subset of the full set of constraints (3.3)–(3.5).[1] These equations of motion can be solved with full generality (see Appendix B.1), but it turns out that the most interesting solutions are the subset with $\mathscr{R}$ constant. These are given by

[1] One should check that the final solution satisfies the equations of motion (3.3)–(3.5), which is indeed the case.

$$
\begin{aligned}
ds_1^2 = &-\left(1-\frac{2G_4M_1}{r}+\frac{CQ^2}{r^2}+\frac{C_3}{r^4}-\frac{G_4\Lambda_1 r^2}{3}\right)dt^2 \\
&+\left(\frac{dr}{\sqrt{1-\frac{2G_4M_1}{r}+\frac{CQ^2}{r^2}+\frac{C_3}{r^4}-\frac{G_4\Lambda_1 r^2}{3}}}+\sqrt{\frac{2G_4M_2}{r}+\frac{C_3}{r^4}+\frac{G_4\Lambda_2 r^2}{3}}dt\right)^2 \\
&+r^2 d\Omega_2,
\end{aligned}
\tag{3.17}
$$

where

$$
\Lambda_1 \equiv \frac{2\lambda}{G_4}\left(qC-1\right), \qquad \Lambda_2 \equiv \frac{\lambda}{G_4}\left(4qC-1-3C^2\right), \tag{3.18}
$$

and Q, C, M_1, M_2, C_3 are integration constants. Moreover, as expected,

$$
\Lambda_1 + \Lambda_2 = \Lambda_C, \tag{3.19}
$$

where Λ_C has been defined in (3.9). This family of solutions is therefore characterized by these five integration constants and the two dimensional constants of the theory (λ, q).

3.1.2.1 Analysis of the Solutions

To understand the physical meaning of the solution (3.17), we perform a coordinate transformation $t \to T = T(t, r)$,

$$
dT = dt - \frac{1}{1-\frac{2G_4M}{r}+\frac{CQ^2}{r^2}-\frac{G_4\Lambda_C r^2}{3}}\sqrt{\frac{\frac{2G_4M_2}{r}+\frac{C_3}{r^4}+\frac{G_4\Lambda_2 r^2}{3}}{1-\frac{2G_4M_1}{r}+\frac{CQ^2}{r^2}+\frac{C_3}{r^4}-\frac{G_4\Lambda_1 r^2}{3}}}dr, \tag{3.20}
$$

where $M \equiv M_1 + M_2$. Somewhat surprisingly, the line element (3.17) becomes recognizable as the Reissner-Nordström-(Anti)-de-Sitter solution with mass M, charge $\sqrt{C}Q$ and cosmological constant Λ_C:

$$
\begin{aligned}
ds_2^2 = &-\left(1-\frac{2G_4M}{r}+\frac{CQ^2}{r^2}-\frac{G_4\Lambda_C r^2}{3}\right)dT^2+\frac{dr^2}{1-\frac{2G_4M}{r}+\frac{CQ^2}{r^2}-\frac{G_4\Lambda_C r^2}{3}} \\
&+r^2 d\Omega_2.
\end{aligned}
\tag{3.21}
$$

Observe the renormalization of the charge, as anticipated in Sect. 3.1.1. One can confirm that this line element is a solution of n-DBI gravity as well.

Geometrically, the solution we have found is nothing but this standard solution of Einstein gravity, written in an unusual set of coordinates that can be thought of as a superposition of Gullstrand-Painlevé,

$$ds_{\mathrm{GP}}^2 = -dt^2 + \left(dr + e^{2g(r)}dt\right)^2 + r^2 d\Omega_2, \tag{3.22}$$

and Schwarzschild coordinates,

$$ds_{\mathrm{Sch}}^2 = -f(r)^2 dt^2 + g(r)^2 dr^2 + r^2 d\Omega_2. \tag{3.23}$$

The coordinate transformation (3.20) is not, however, a foliation preserving diffeomorphism. Thus, in n-DBI gravity, (3.17) and (3.21) describe the same solution if and only if $M_2 = C_3 = \Lambda_2 = 0$. Otherwise they are two distinct solutions with different physical invariants (discussed below) which *happen to* be mapped by a non-foliation preserving diffeomorphism. Namely, we can schematically describe the situation as follows:

$$ds_1 \overset{\mathrm{Diff}}{\rightarrow} ds_2 \quad \Rightarrow \quad ds_1 \text{ and } ds_2 \text{ are NOT physically equivalent},$$

$$ds_1 \overset{\mathrm{Diff}\,\mathcal{F}}{\rightarrow} ds_2 \quad (\text{iff. } M_2 = C_3 = \Lambda_2 = 0) \quad \Rightarrow \quad ds_1 \text{ and } ds_2 \text{ are physically equivalent.}$$

In the above, "$\overset{\mathrm{Diff}}{\rightarrow}$" stands for the general (non-foliation preserving) map (3.20), and especially when the map (3.20) becomes the foliation preserving diffeomorphism, it is written as "$\overset{\mathrm{Diff}\,\mathcal{F}}{\rightarrow}$". Thereby, one can notice that in n-DBI gravity the constants (M_2, C_3, Λ_2) are newly induced by the symmetry breaking. Thus, the constants (M_2, C_3, Λ_2) should specify the foliation because in n-DBI gravity there is no DOF to deform the foliation as opposed to general relativity, and so a specific choice of the foliation breaks the full diffeomorphism down to the foliation preserving one. To check the statement above, remember the definition of the extrinsic curvature:

$$K_{\mu\nu} = \frac{1}{2}\pounds_n h_{\mu\nu}. \tag{3.24}$$

We introduce its traceless part σ_{ij} and the trace part θ as well:

$$\sigma_{ij} = K_{ij} - \frac{1}{3}K h_{ij}, \qquad \theta = K, \tag{3.25}$$

where σ_{ij} and θ are called *shear* and *expansion* of the congruence of time-like curves tangent to n, resectively. As is clear from the definition, the extrinsic curvature measures variations of the foliation along the normal direction n: the shape (embedding) of the foliation is determined by the extrinsic curvature. Let us take a look at the scalar (gauge-invariant) quantities constructed by the extrinsic curvature:

$$\theta = -\frac{3\left(G_4 M_2 + G_4 \Lambda_2 r^3\right)}{\sqrt{C_3 + 2G_4 M_2 r^3 + G_4 \Lambda_2 r^6}},$$

$$\sigma_{ij}\sigma^{ij} = 6\left[\frac{C_3 + G_4 M_2 r^3}{r^3\sqrt{C_3 + 2G_4 M_2 r^3 + G_4 \Lambda_2 r^6}}\right]^2; \tag{3.26}$$

we also have

$$K_{ij}K^{ij} - K^2 = 6\left(\frac{C_3}{r^6} - G_4\Lambda_2\right). \tag{3.27}$$

It is manifest that M_2, Λ_2 and C_3 enter in these scalar invariants. One can understand that M_2 and Λ_2 dress the mass of the black hole and the cosmological constant, respectively. However, the physical meaning of C_3 is still unclear. To uncover its meaning, we set all coefficients zero except for C_3 in the line element (3.17):

$$ds_1^2 = -\left(1 + \frac{C_3}{r^4}\right)dt^2 + \left(\frac{dr}{\sqrt{1+\frac{C_3}{r^4}}} + \sqrt{\frac{C_3}{r^4}}dt\right)^2 + r^2 d\Omega_2. \tag{3.28}$$

The mass, cosmological constant and charge of the black hole have been set to zero in (3.28). Nevertheless, one can observe the singularity at $r = 0$, and besides it can not be gauged away. We dub this singularity as *shearing singularity* [1]. This is because when setting $M_2 = \Lambda_2 = 0$ in the scalar quantities made of the extrinsic curvature, one finds that the shear only survives as

$$K_{ij}K^{ij} = \sigma_{ij}\sigma^{ij} = \frac{6C_3}{r^6}, \quad \theta = 0. \tag{3.29}$$

We summarize the consequence. We found the solution ds_1^2 with 5 unknown constants $\{Q, C, M_1, M_2, M_3, C_3\}$. Performing the non-foliation preserving diffeomorphism, the C_3 was removed and we obtained ds_2^2. The ds_2^2 was turned out to be a solution in n-DBI gravity, the Reissner-Nordström-(Anti)-de-Sitter solution. However, this transformation is not a foliation preserving one. This means that the C_3 is not a gauge artifact but a shearing singularity. As usual in Einstein gravity, one can invoke smoothness to rule out some solutions as unphysical. For instance smoothness of the constant (Riemann) curvature spaces ($M_1 = M_2 = Q = 0$), requires $C_3 = 0$ to avoid the shearing singularity at $r = 0$.

Notice that there is, for example, no Reissner-Nordström (A)dS solution in Gullstrand-Painlevé coordinates. If the symmetry group of n-DBI gravity was the set of general coordinate transformations, we would have found such a solution. In other words, the breakdown of the symmetry to foliation preserving diffeomorphisms is explicitly reflected in the form of the solutions (3.17).

3.1.2.2 Asymptotic Behavior

Asymptotically ($r \to \infty$) the solution (3.17) becomes a constant curvature space:

$$R_{\mu\nu\alpha\beta} = \frac{G_4\Lambda_C}{3}\left(g_{\mu\alpha}g_{\nu\beta} - g_{\mu\beta}g_{\nu\alpha}\right). \tag{3.30}$$

With appropriate choices of C one may set either $\Lambda_1 = 0$ or $\Lambda_2 = 0$, keeping the other non-vanishing. In both cases one recognizes de Sitter space: either written in Painlevé-Gullstrand coordinates (with cosmological constant Λ_2), or written in static coordinates (with cosmological constant Λ_1). In the latter case, Anti-de-Sitter space may also occur, written in global coordinates. Keeping both Λ_1 and Λ_2 non-vanishing one has an unusual slicing of constant curvature spaces. This can represent de Sitter space-time, Anti-de Sitter space-time or Minkowski space, depending on the sign of the total cosmological constant $\Lambda_1 + \Lambda_2$. Indeed, the integration constant C controls the magnitude of the cosmological constant:

$$\begin{aligned} C &\in \left]q - \sqrt{q^2 - 1}, q + \sqrt{q^2 - 1}\right[, && \text{de Sitter} \\ C &= q \pm \sqrt{q^2 - 1}, && \text{Minkowski} \\ C &\text{ otherwise} && \text{AdS.} \end{aligned}$$

dS and Minkowski space solutions can only exist if $q \geq 1$.

3.1.3 Discussion

In [1] we have explored some further properties of n-DBI gravity, beyond those studied in [2], which focused on cosmology.

A crucial property of the theory is the existence of an everywhere time-like vector field n. We have assumed it to be hypersurface orthogonal—which is expressed by the relation we have chosen between n and the ADM quantities—albeit this condition could be dropped and an even more general framework considered. The existence and role played by n is reminiscent of Einstein-æther theory (see [4] for a review).

It follows that the symmetry group of the theory is that which preserves n and therefore the foliation defined by it, $\mathscr{F}$. This group, denoted by $\text{Diff}_{\mathscr{F}}(\mathscr{M})$, is the group of foliation preserving diffeomorphisms, and it is therefore smaller than general coordinate transformations; it is the group that leaves invariant the equations of motion. This means that a non-foliation preserving diffeomorphism applied to a solution of n-DBI gravity is *not*, in general, a solution of n-DBI gravity. Exceptionally, however, this may happen and a non-foliation preserving diffeomorphism may map two solutions. These should be regarded, however, as physically distinct solutions, perhaps in the same orbit of a larger symmetry group, in the same spirit of many duality symmetries or solution generating techniques that have been considered in the context of supergravity or string theory. In n-DBI gravity it is unclear, at the moment, if such larger symmetry group exists, but an explicit example of a non-foliation preserving mapping (inequivalent) solutions was provided by (3.20). The solutions are, of course, isometric; in this particular example they are the standard Reissner-Nordström-(A)dS geometry in two different coordinate systems. Observe, however, the non-trivial dynamics of the theory, where the mass and the cosmological constant can in effect be split between two slicings but not the charge.

The fact that the spherically symmetric solutions of n-DBI gravity minimally coupled to a Maxwell field contain precisely the Reissner-Nordström geometry (with or without a cosmological constant) is remarkable and, as far as we are aware, unparalleled, within theories of gravity with higher curvature terms. This leads to the natural question of how generic is it that Einstein gravity solutions are solutions of n-DBI gravity (with the same matter content)? Following the theorem and corollary presented in Sect. 3.1.1 this question can be recast very objectively as the existence of a foliation with a specific property. How generically can such foliation be found? Can it be found for the Kerr solution?

Finally, the ansatz compatible with spherical symmetry in n-DBI gravity has more degrees of freedom than in Einstein gravity. It will be quite interesting to see if, even in vacuum, such ansatz can accommodate a non-trivial time dependence, prohibited in Einstein gravity by Birkhoff's theorem.

3.2 Scalar Graviton in n-DBI Gravity

n-DBI gravity is a gravitational theory which yields near de Sitter inflation spontaneously at the cost of breaking Lorentz invariance by a preferred choice of foliation. In [3], we have shown that this breakdown endows n-DBI gravity with one extra physical gravitational degree of freedom: a *scalar graviton*. Its existence is established by Dirac's theory of constrained systems. Firstly, a general analysis is made in the canonical formalism, using ADM variables. It is useful to introduce an auxiliary scalar field, which allows recasting n-DBI gravity in an Einstein-Hilbert form but in a Jordan frame. Identifying the constraints and their classes we confirm the existence of an extra degree of freedom in the full theory, besides the two usual tensorial modes of the graviton. Then, studying scalar perturbations around Minkowski space-time, we show that there exists one scalar degree of freedom and identify it in terms of the metric perturbations. We then argue that, unlike the case of (the original proposal for) Hořava-Lifshitz (HL) gravity [5], there is no evidence that the extra degree of freedom originates pathologies, such as vanishing lapse, instabilities and strong self-coupling at low energy scales.

3.2.1 *Nailing Scalar Graviton*

We will consider the Hamiltonian formulation of the full n-DBI gravity theory and confirm the existence of one extra degree of freedom. Since in the n-DBI action the time derivative is of first order at most, one can work on the Hamiltonian formalism. The n-DBI Lagrangian, L_{nDBI}, is given by Eq. (3.1). The square root makes the analysis cumbersome, and we find it more convenient to work in the linearized form with the auxiliary field e (1.133):

$$L^{e}_{\text{nDBI}} = -\frac{1}{\kappa}\int d^3x\sqrt{-g}\, e\left(R + K_{ij}K^{ij} - K^2 - 2N^{-1}\Delta N - 2G_4\Lambda_C(e)\right), \tag{3.31}$$

where $\kappa \equiv 16\pi G_4$. In passing to the Hamiltonian formalism, we set the notation for the canonical conjugate momenta as

$$L^{e}_{\text{nDBI}}\left((h_{ij},\dot{h}_{ij}),(N,\dot{N}),(N_i,\dot{N}_i),(e,\dot{e})\right) \to H^{e}_{\text{nDBI}}\left((h_{ij},p^{ij}),(N,p_N),(N_i,p^i_{\mathbf{N}}),(e,p_e)\right). \tag{3.32}$$

However, the time derivatives of N and N_i are absent as in general relativity, and so is the time derivative of e. Thus we have the primary constraints,

$$\Phi_1 \equiv p_N = 0, \quad \Phi^i_2 \equiv p^i_{\mathbf{N}} = 0, \quad \Phi_3 \equiv p_e = 0. \tag{3.33}$$

Denoting the Lagrangian density by $\mathscr{L}^{e}_{\text{nDBI}}$, the Hamiltonian density is given by

$$\begin{aligned}\mathscr{H}^{e(0)}_{\text{nDBI}} &\equiv p^{ij}\dot{h}_{ij} - \mathscr{L}^{e}_{\text{nDBI}}\\ &= \sqrt{h}N_j\left(-\frac{2}{\sqrt{h}}\nabla_i p^{ij}\right) + \frac{\sqrt{h}N}{\kappa}\left[-\frac{\kappa^2}{eh}\left(p^{ij}p_{ij} - \frac{1}{2}p^2\right) + e\left(R - 2G_4\Lambda_C(e)\right)\right]\\ &\quad - \frac{2}{\kappa}\sqrt{h}\,(e\Delta N),\end{aligned} \tag{3.34}$$

where

$$p^{ij} \equiv \frac{\delta L^{e}_{\text{nDBI}}}{\delta \dot{h}_{ij}} = -\frac{\sqrt{h}}{\kappa}(K^{ij} - h^{ij}K)e. \tag{3.35}$$

The time flows of the constraints are generated by the extended Hamiltonian density

$$\mathscr{H}^{e(1)}_{\text{DBI}} = \mathscr{H}^{e(0)}_{\text{DBI}} + \lambda_1\Phi_1 + \lambda_{2i}\Phi^i_2 + \lambda_3\Phi_3, \tag{3.36}$$

where λ_1, λ_{2i} and λ_3 are the Lagrange multipliers. Thus the primary constraints evolve in time as

$$\begin{aligned}\dot{\Phi}_1(x) &= \int d^3y\{\mathscr{H}^{e(0)}_{\text{DBI}}(y), \Phi_1(x)\} + \sum_{a=1,3}\int d^3y\{\Phi_a(y), \Phi_1(x)\}\lambda_a\\ &\quad + \int d^3y\{\Phi^i_2(y), \Phi_1(x)\}\lambda_{2i}\\ &= -\frac{\sqrt{h}}{\kappa}\left[-\frac{\kappa^2}{eh}\left(p^{ij}p_{ij} - \frac{1}{2}p^2\right) + e\left(R - 2G_4\Lambda_C(e)\right) - 2\Delta e\right] \equiv \Phi_4(x),\end{aligned} \tag{3.37}$$

$$
\begin{aligned}
\dot{\Phi}_2^i(x) &= \int d^3y\{\mathscr{H}_{\mathrm{DBI}}^{e(0)}(y), \Phi_2^i(x)\} + \sum_{a=1,3} \int d^3y\{\Phi_a(y), \Phi_2^i(x)\}\lambda_a \\
&\quad + \int d^3y\{\Phi_2^j(y), \Phi_2^i(x)\}\lambda_{2j} \\
&= 2\nabla_j p^{ij} \equiv \Phi_5^i(x),
\end{aligned} \tag{3.38}
$$

$$
\begin{aligned}
\dot{\Phi}_3(x) &= \int d^3y\{\mathscr{H}_{\mathrm{DBI}}^{e(0)}(y), \Phi_3(x)\} + \sum_{a=1,3} \int d^3y\{\Phi_a(y), \Phi_3(x)\}\lambda_a \\
&\quad + \int d^3y\{\Phi_2^i(y), \Phi_3(x)\}\lambda_{2i} \\
&= \frac{\sqrt{h}N}{\kappa e}\left[-\frac{\kappa^2}{eh}\left(p^{ij}p_{ij} - \frac{1}{2}p^2\right) - e\left(R - 2G_4\Lambda_C(e)\right) - \frac{12\lambda}{G_4}\left(q - \frac{1}{e}\right)\right] \\
&\quad + \frac{2\sqrt{h}}{\kappa}\Delta N \equiv \Phi_6(x).
\end{aligned} \tag{3.39}
$$

Therefore, in addition to the primary constraints (3.33), we have the secondary constraints

$$
\Phi_4 = \dot{p}_N \approx 0\,, \quad \Phi_5^i = \dot{p}_{\mathbf{N}}^i \approx 0\,, \quad \Phi_6 = \dot{p}_e \approx 0\,. \tag{3.40}
$$

Finally, the time flows of the secondary constraints do not yield any further constraints, as shown in Appendix B.2.

It is easy to understand the physical meaning of the above constraints. Firstly, we can solve $\Phi_6 = 0$ for e and obtain

$$
e = \pm\left(1 + \frac{G_4}{6\lambda}\mathscr{R}\right)^{-1/2}, \tag{3.41}
$$

where we defined $\mathscr{R} \equiv {}^{(4)}R + \mathscr{K}$. Choosing the positive sign and plugging it into Φ_4, we find

$$
\begin{aligned}
\Phi_4 &= -\frac{\kappa}{2}\Bigg[-\tfrac{3\lambda}{4\pi G_4^2}q + \tfrac{1}{\sqrt{h}}\sqrt{A(h,N)B(h,p)} \\
&\quad + \tfrac{1}{\sqrt{h}}\tfrac{\Delta N}{N}\sqrt{\tfrac{B(h,p)}{A(h,N)}} - \Delta\left(\tfrac{1}{\sqrt{h}}\sqrt{\tfrac{B(h,p)}{A(h,N)}}\right)\Bigg],
\end{aligned} \tag{3.42}
$$

where

$$
A(h,N) = \frac{6\lambda}{G_4} + R - 2N^{-1}\Delta N, \quad B(h,p) = 2p^2 - 4p^{ij}p_{ij} + \frac{3\lambda h}{32\pi^2 G_4^3}. \tag{3.43}
$$

In terms of the Lagrangian variables, the constraint $\Phi_4 = 0$ yields

$$0 = \frac{1 + \frac{G_4}{6\lambda}(R - N^{-1}\Delta N)}{\sqrt{1 + \frac{G_4}{6\lambda}\mathcal{R}}} - q - \frac{G_4}{6\lambda}\Delta\left(1 + \frac{G_4}{6\lambda}\mathcal{R}\right)^{-1/2}. \tag{3.44}$$

This is nothing but the Hamiltonian constraint in [1]. Similarly, plugging (3.41) into (3.35), the constraints $\Phi_5^i = 0$ yield

$$\nabla_j\left(\frac{K^{ji} - h^{ji}K}{\sqrt{1 + \frac{G_N}{6\lambda}\mathcal{R}}}\right) = 0. \tag{3.45}$$

These are the momentum constraints in [1]. We, however, note that it is more appropriate to regard the following linear combination:

$$\tilde{\Phi}_{5j} = \Phi_{5j} - \Phi_1\partial_j N - \Phi_3\partial_j e, \tag{3.46}$$

as the momentum constraints. This is because these are the constraints that generate the spatial diffeomorphisms for the phase-space variables, h_{ij}, p^{ij}, N and e, rather than Φ_{5j} (see Appendix B.2). By an explicit computation, one can show that Φ_{2j} and $\tilde{\Phi}_{5j}$ are first-class constraints, forming the constraint algebra (B.66)–(B.71), and the rest are second class (see Appendix B.2 for details). Hence, the constraints are classified as

$$\begin{aligned}
&\#(\text{phase}-\text{space variables}) = \#(h_{ij}, p^{ij}, N, p_N, N_i, p_{\mathrm{N}}^i, e, p_e) = 2(6+1+3+1) = 22,\\
&\#(\text{2nd}-\text{class constraints}) = \#(\Phi_1, \Phi_3, \Phi_4, \Phi_6) = 1+1+1+1 = 4,\\
&\#(\text{1st}-\text{class constraints}) = \#(\Phi_{2j}, \tilde{\Phi}_{5j}) = 3+3 = 6.
\end{aligned}$$

Consequently, the number of physical degrees of freedom (DOF) in n-DBI gravity reads

$$\text{DOF of graviton} = \frac{1}{2}(22 - 4 - 2\times 6) = 2 + 1. \tag{3.47}$$

Indeed, as advertised, we find an extra degree of freedom as compared to general relativity. This is nothing but a *scalar graviton*.

3.2.2 Identity of Scalar Graviton

Having established the existence of the scalar graviton, we wish to identify it in terms of the metric perturbations from the flat background. The metric has ten components: the lapse N, the shift N^i, and the spatial metric h_{ij}. These can be decomposed into 4 scalars (n, B, ψ, E), 2 transverse vectors $(A_i, \tilde{A}_i)$, 1 transverse traceless tensor $\tilde{h}_{ij}$

as follows (see e.g. [6]):

$$N = 1 + n, \qquad N_i = \nabla_i B + A_i,$$
$$h_{ij} = \delta_{ij} - 2\left(\delta_{ij} - \frac{\nabla_i \nabla_j}{\Delta}\right)\psi - 2\frac{\nabla_i \nabla_j}{\Delta}E + \left(\nabla_i \tilde{A}_j + \nabla_j \tilde{A}_i\right) + \tilde{h}_{ij}, \tag{3.48}$$

where the transversality and traceless conditions are imposed,

$$\nabla^i A_i = \nabla^i \tilde{A}_i = \nabla^i \tilde{h}_{ij} = \tilde{h}^i_i = 0. \tag{3.49}$$

We now expand the n-DBI gravity action (3.1) to quadratic order around flat spacetime, setting $q = 1$. Thanks to conditions (3.49), scalar perturbations and vector-tensor perturbations decouple from each other. The quadratic Lagrangian density for the scalar fields yields, up to total divergences,

$$4\pi G_4 \mathscr{L}_{\text{scalar}} = 2\dot{\psi}^2 + 4\dot{\psi}\left(\dot{E} + \Delta B\right) + (2\psi - 4n)\,\Delta\psi + \frac{G_4}{6\lambda}\left(2\Delta\psi - \Delta n\right)^2. \tag{3.50}$$

This essentially corresponds to the $\lambda = 1$ case of [7], and the scalar graviton in this model is qualitatively different in its character from the one discussed there. In order to see the scalar degree of freedom in the flat space, we shall apply Dirac's theory of constrained systems. The Hamiltonian density of this effective scalar theory reads[2]

$$\mathscr{H}^{(0)}_{\text{scalar}} = -\frac{1}{8}p_E^2 + \frac{1}{4}p_E p_\psi - p_E \Delta B - (2\psi - 4n)\,\Delta\psi - \frac{G_4}{6\lambda}\left(2\Delta\psi - \Delta n\right)^2, \tag{3.51}$$

where p_E and p_ψ are the conjugate momenta of E and ψ, respectively. Since the Lagrangian density (3.50) does not contain the time derivative of n or B, their conjugate momenta become the following primary constraints:

$$\Phi_1 \equiv p_n = 0, \qquad \Phi_2 \equiv p_B = 0. \tag{3.52}$$

Thus the dynamics of this system is governed by the Hamiltonian density

$$\mathscr{H}^{(1)}_{\text{scalar}} = \mathscr{H}^{(0)}_{\text{scalar}} + \lambda_1 \Phi_1 + \lambda_2 \Phi_2, \tag{3.53}$$

where λ_1 and λ_2 are Lagrange multipliers. The consistency requires the time flows of the primary constraints to vanish:

[2] For convenience, we have rescaled the scalar fields by the factor of $(4\pi G_4)^{1/2}$.

$$\Phi_4 \equiv \dot{\Phi}_1 = 4\Delta\psi + \frac{G_4}{6\lambda}\left(4\Delta^2\psi - 2\Delta^2 n\right) \approx 0, \qquad \Phi_5 \equiv \dot{\Phi}_2 = -\Delta p_E \approx 0. \tag{3.54}$$

The former corresponds to the Hamiltonian constraint and the latter to the momentum constraint. The time flows of these secondary constraints do not yield new constraints. Among these four constraints, Φ_2 and Φ_5 commute with all the constraints and are therefore first class, whereas Φ_1 and Φ_4 are second class. Hence the physical degrees of freedom of this system are counted as $(2 \times 4 - 2 \times 2 - 2)/2 = 1$. In contrast, in general relativity (herein we dub as GR) where $\lambda \to \infty$, all the constraints are first class and the counting becomes $(2\times 4 - 2\times 2 - 2\times 2)/2 = 0$. In other words, GR has two pairs of first class constraints associated with the full diffeomorphism, whereas n-DBI has only one pair reflecting that only the foliation preserving diffeomorphism is preserved. Thus, in n-DBI gravity there remains one physical scalar degree of freedom that cannot be gauged away. We will elaborate on this point in a moment.

Note that it is the presence of the lapse term $(\Delta n)^2 \sim (N^{-1}\Delta N)^2$ in the Hamiltonian that leaves one physical scalar degree of freedom, as opposed to $1/2$ in Hořava-Lifshitz (HL) gravity. [3] This is similar to the consistent extension of HL gravity [7].

We shall clarify how the scalar mode becomes physical as a direct consequence of the breakdown of full diffeomorphism invariance to foliation preserving one. For this purpose, we first note that the scalar field theory (3.50) has the foliation preserving gauge symmetry:

$$\psi \to \psi, \tag{3.55}$$

$$B \to B + \dot{L} - T, \tag{3.56}$$

$$E \to E - \Delta L, \tag{3.57}$$

$$n \to n + \dot{T}, \tag{3.58}$$

where T and L are related to infinitesimal coordinate transformations by $\xi^0 = T$ and $\xi^i = \nabla^i L$. In n-DBI gravity, T is a function of time only, whereas L is a function of both space and time. In GR, T also becomes a function of both space and time. Note that the gauge invariant quantities are

[3] In non-projectable HL gravity, the lapse N is a Lagrange multiplier as in GR, but the Hamiltonian constraint is second class. Moreover, the time flow of the Hamiltonian constraint yields an additional constraint that depends on the lapse N. Thus, there are 3 second class constraints; the conjugate momentum p_N of the lapse, the Hamiltonian constraint $\mathcal{H}$, and its time flow $\dot{\mathcal{H}}$. Together with 6 first class constraints, the number of degrees of freedom is counted as $(2\times 10 - 2\times 6 - 3)/2 = 2 + 1/2$. Note, however, that linear perturbations about flat space-time yield a misleading result. There appear four second and two first class constraints, implying incorrectly that the number of scalar degrees of freedom is zero. In projectable HL gravity, there is an additional primary constraint $\partial_i N = 0$ on top of $p_N = 0$. In this case, the time flow of $p_N = 0$ does not yield the Hamiltonian constraint. Instead, it determines the Lagrange multiplier of $\partial_i N$. Hence p_N and $\partial_i N$ are the only second class constraints, and the total physical degrees of freedom is $2 + 1$.

$$\{\psi, \ddot{E} + \Delta\dot{B} + \Delta n\} \qquad \text{for both GR and } n\text{-DBI}, \tag{3.59}$$

$$\{\dot{E} + \Delta B, \partial_i n\} \qquad \text{for } n\text{-DBI only}. \tag{3.60}$$

As it is clear from the above counting of degrees of freedom, the extra scalar graviton exists in n-DBI gravity, because the constraints Φ_1 and Φ_4 are second class, whereas in GR they are first class and generate the gauge transformations

$$\left\{\int d^3y \zeta_1(y)\Phi_1(y), n(x)\right\} = \zeta_1(x), \qquad \left\{\int d^3y \zeta_2(y)\Phi_4(y), p_\psi(x)\right\} = -4\Delta\zeta_2(x). \tag{3.61}$$

Comparing this with the foliation preserving diffeomorphism (3.55)–(3.58), we find that

$$\zeta_1 = \dot{T}, \qquad \zeta_2 = T, \tag{3.62}$$

where we used $p_\psi = 4(\dot{\psi} + \dot{E} + \Delta B)$. In n-DBI gravity, as we stressed above, T is a function of time only. Accordingly, the constraints Φ_1 and Φ_4 become second class and are not considered as generators of gauge transformations. This implies that the scalar graviton should involve the lapse and/or the shift whose gauge transformations are generated by T. This nicely fits the expectation that the scalar graviton must have something to do with the foliation structure which is specified by the lapse and the shift. Indeed, the equations of motion,

$$\ddot{\psi} = 0, \tag{3.63}$$

$$\Delta\dot{\psi} = 0, \tag{3.64}$$

$$\ddot{E} + \Delta\dot{B} + \Delta n + \Delta\psi = 0, \tag{3.65}$$

$$\Delta\psi = \frac{G_N}{6\lambda}\left(\frac{1}{2}\Delta^2 n - \Delta^2\psi\right), \tag{3.66}$$

have in the $E = 0$ gauge the general solution[4]

$$B(t, x) = B_0(x) + B_1(x)t, \tag{3.67}$$

$$n(t, x) = -\, B_1(x) - \psi_0(x), \tag{3.68}$$

$$\psi(t, x) = \psi_0(x), \tag{3.69}$$

with $\psi_0(x)$ related to $B_1(x)$ by

4 A more common gauge is to set the shift $B = 0$. We can go from the $E = 0$ to the $B = 0$ gauge by choosing the gauge parameter $L(t, x) = -B_0(x)t - \frac{1}{2}B_1(x)t^2$. This yields the conformal mode $E(t, x) = \Delta B_0(x)t + \frac{1}{2}\Delta B_1(x)t^2$. Note that in either gauge the lapse n alone only accounts for a half degree of freedom of the scalar graviton.

$$\psi_0(x) = -\frac{G_N}{6\lambda}\left(\Delta B_1(x) + \frac{3}{2}\Delta\psi_0(x)\right). \tag{3.70}$$

Note that we have imposed the boundary condition that all the fields must fall off at spatial infinity. In other words, the functions $f(x)$'s appearing in the most general solution and obeying the Laplace equation $\Delta f(x) = 0$ are unphysical and set to zero. In the Hamiltonian system, a degree of freedom is the freedom to choose a pair of initial data for the time evolution in the phase space. We have found exactly one degree of freedom, *i.e.*, the initial data specified by a pair of arbitrary functions of space, $(B_0(x), B_1(x))$. In GR, these could have been gauged away by choosing the gauge parameter $T(t,x) = B(t,x)$ (and the Hamiltonian constraint would have enforced $\psi_0(x) = 0$). Put differently, in n-DBI gravity the scalar mode is the broken gauge degree of freedom $T(t,x)$ which obeys, by taking the time derivative of the the Hamiltonian constraint (3.66),

$$\Delta^2 \ddot{T}(t,x) = 0, \tag{3.71}$$

where we have used $n(t,x) = -\dot{T}(t,x) - \psi_0(x)$ and $\psi(t,x) = \psi_0(x)$. A few remarks are in order: firstly, this is a key equation, despite its extremely simple appearance, and notably originates from the nonlinear lapse term $(\Delta n)^2 \sim (N^{-1}\Delta N)^2$ in the Hamiltonian; secondly, the Stückelberg field satisfies exactly the same equation and thus can be identified with the broken gauge degree of freedom $T(t,x)$.

Observe that the frequency of the scalar mode is $\omega = 0$, and thus this is more a zero mode than a propagating particle mode. Nonetheless, this is the physical degree of freedom of our Hamiltonian system. We will postpone the interpretation of this result for the moment and discuss it in Sect. 3.2.4.

3.2.3 *Potential Pathologies*

We shall now address some potential pathologies associated with the existence of the scalar graviton, which are known to afflict HL gravity.

3.2.3.1 Vanishing Lapse

One of the most serious problems of HL gravity is the absence of dynamics discussed in [8]. The problem is that the time flow of the Hamiltonian constraint yields an independent constraint on the lapse, which, as it turns out, requires the lapse to identically vanish in asymptotically flat space-times, implying that HL gravity does not have any dynamics. Henneaux et al. [8] emphasize that this problem is intimately related to the fact that in HL gravity the Hamiltonian constraint is second class. In n-DBI gravity, the Hamiltonian constraint is also second class. This raises the concern that this model might also lack dynamics. However, this problem is evaded in a similar way as the consistent extension of HL gravity does. Namely, the time flow of the Hamiltonian constraint, given by (B.72) in the full theory, does not yield an additional constraint unlike HL gravity; rather it determines the Lagrange multiplier λ_3, since the Hamiltonian constraint Φ_4 depends on the auxiliary field e (and implicitly on the lapse N through e by solving $\Phi_6 = 0$) and thus $\{\Phi_3, \Phi_4\} \neq 0$. This stems from the *nonlinear* lapse dependence in the n-DBI action and can be seen more clearly in the linearized theory in Sect. 3.2.1[5]: the time flow of the Hamiltonian constraint Φ_4 is generated by its commutator with the Hamiltonian density (3.53). Because of the lapse term $\Delta^2 n$ in the Hamiltonian constraint descended from the nonlinear lapse term $(\Delta n)^2 \sim (N^{-1}\Delta N)^2$ in the action, the coefficient $\{\Phi_1, \Phi_4\}$ of the Lagrange multiplier λ_1 is non-vanishing. Thus, such time flow determines λ_1 rather than imposing an extra constraint on the lapse.

To summarize, the nonlinear lapse dependence intrinsic to n-DBI gravity rescues the model from the absence of dynamics.

3.2.3.2 Short Distance Instability

In [9] it was found that the scalar graviton in HL gravity develops an exponential time growth at short distances in generic space-times in the linearized approximation. This suggests the presence of a universal short distance instability in HL gravity. Such problem, however, should be considered with some caution. The exponentially growing mode involves the lapse; but the lapse in asymptotically flat space-times is forced to vanish everywhere [8], as mentioned in Sect. 3.2.3.1. If the lapse must collapse everywhere, there would be no way to develop an exponentially growing mode involving the lapse. Thus there appears to be a contradiction between these two issues. This might imply that the unstable scalar mode found in [9] fails to obey the boundary condition at spatial infinity, when extended to long distances. In any case, as we have just shown, the lapse does not vanish in our model and so it is sensible and important to study whether the scalar mode leads to any sort of instability or not.

We shall now show that the analysis that reveals an exponential time growth at short scales in HL gravity, does not allow the same conclusion when applied to the scalar graviton in n-DBI.We begin by recasting the equations of motion of n-DBI gravity:

$$(\pounds_t - \pounds_N)h_{ij} - 2NK_{ij} = 0, \tag{3.72}$$

$$R + K^2 - K_{ij}K^{ij} = 2G_4\Lambda_C(e) + 2e^{-1}\Delta e, \tag{3.73}$$

$$\nabla^j \left(K_{ij} - h_{ij}K\right) = -e^{-1}\nabla^j e \left(K_{ij} - h_{ij}K\right), \tag{3.74}$$

[5] In the full theory, the nonlinear lapse dependence is somewhat obscured by the introduction of the auxiliary field e which linearizes the lapse dependence.

$$
\begin{aligned}
(\pounds_t - \pounds_N)\left(K^{ij} - h^{ij}K\right) + N\left(R^{ij} - KK^{ij} + 2K^{il}K^j{}_l - h^{ij}K_{mn}K^{mn}\right) \\
- \nabla^i\nabla^j N + h^{ij}\Delta N \\
= - e^{-1}(\pounds_t - \pounds_N)e\left(K^{ij} - h^{ij}K\right) \\
+ e^{-1}N\left(\nabla^i\nabla^j - h^{ij}(\nabla_l \ln N)\nabla^l\right)e,
\end{aligned}
\tag{3.75}
$$

where $e = \left(1 + \frac{G_4}{6\lambda}\mathscr{R}\right)^{-\frac{1}{2}}$ and the cosmological constant $\Lambda_C(e) = \frac{3\lambda}{G_4^2}\left(\frac{2q}{e} - 1 - \frac{1}{e^2}\right)$. The first equation defines the extrinsic curvature. The second and third equations are the Hamiltonian and momentum constraints, respectively. The last equation is the evolution equation [1]. In this form, the R.H.S. of (3.73)–(3.75) represent the modification from GR. In fact, when the auxiliary field e is constant, the R.H.S. vanish except for the cosmological constant in the Hamiltonian constraint (3.73), and these equations become those of GR, as remarked in Sect. 3.1.1. As we have seen in (3.71), the time-derivative of the Hamiltonian constraint succinctly captures the scalar mode dynamics, and it can be written as

$$
\Delta\dot{e} - KN\Delta e - \left(N\partial^i K + 2K\partial^i N\right)\partial_i e - N^{-1}\Delta N\dot{e} = 0. \tag{3.76}
$$

We now consider small perturbations around a generic background in the gauge $N^i = 0$:

$$
h_{ij} = \bar{h}_{ij} + \gamma_{ij}, \tag{3.77}
$$

$$
K^{ij} = \bar{K}^{ij} + \kappa^{ij}, \tag{3.78}
$$

$$
N = \bar{N} + n. \tag{3.79}
$$

The details of the following analysis can be found in Appendix B.3.1. We assume that the background space-time varies over the characteristic scale L; $\bar{R}_{ij} \sim 1/L^2$, $\bar{K}_{ij} \sim 1/L$, $\bar{\mathscr{R}} \sim 1/L^2$, $\partial \ln \bar{N} \sim 1/L$, and $\bar{e} \sim 1 + O(1/L^2)$. Since we are interested in a potential universal short distance instability, we focus on scales much shorter than L;

$$
\omega, p \gg 1/L\,, \tag{3.80}
$$

where the space-time is nearly flat. We can then neglect terms with space-time derivatives of the background fields relative to those of the perturbed fields. Assuming the Fourier decompositions

$$
\gamma_{ij}, \kappa^{ij}, n \sim e^{-i\omega t + ip\cdot x}, \tag{3.81}
$$

the perturbations of (3.72)–(3.75) and (3.76) yield (B.82)–(B.85) and (B.90), respectively. We then find that for the scalar mode to be present

$$\omega \sim \bar{N}p \sim i\left(\bar{N}\bar{K} + \partial\bar{N}\right) \sim \frac{i}{L}. \tag{3.82}$$

At first sight, it might seem to imply the existence of an exponentially growing mode, suggesting an instability. This relation, however, violates the validity of our approximation (3.80) and thus cannot be satisfied. We conclude that this analysis does not reveal a potential universal short distance instability as discussed in the HL case [9]. Note that this also implies that there is no scalar mode of the type $(n, \gamma) \sim (n(\omega, p), \gamma(\omega, p))\, e^{i\omega t + ip\cdot x}$ at short distances.[6] The same conclusion can be reached by an alternative analysis using the Stückelberg field ϕ. We leave the details of the analysis in Appendix B.3.2. The equation of motion for ϕ can be obtained from the action (1.140). The preferred choice of time-like vector $\mathbf{n} = (-N, 0, 0, 0)$ corresponds to $\phi = t$. Indeed, the equation of motion (B.96) for ϕ reduces, when $\phi = t$, to the time derivative of the Hamiltonian constraint (3.76). We now consider perturbations in unitary gauge $\phi = t + \varphi$ and $N^i = 0$, expanding (3.76) to linear order in φ. To leading order, we find

$$\Delta^2\ddot{\varphi} - (\bar{N}\bar{K} - \partial_t \ln \bar{N})\Delta^2\dot{\varphi} + 2\partial_i \ln \bar{N}\partial^i \Delta\ddot{\varphi} = 0. \tag{3.83}$$

This indeed yields (3.82) found in the previous analysis.

We conclude by remarking that, as advertised, $\varphi(t, x)$ can be identified with the broken gauge parameter $T(t, x)$. Since in flat space-time (3.83) yields $\Delta^2\ddot{\varphi} = 0$, we obtain precisely the scalar graviton equation (3.71) whose origin can be traced to the nonlinear lapse term (observe also that (3.83) is the time derivative of the linearized Hamiltonian constraint). This provides a consistency check of our computations.

3.2.3.3 Strong Coupling

Naively, n-DBI gravity admits the GR limit by tuning the parameter $\lambda \to \infty$ and $q \to 1$ with $\lambda(q-1)$ fixed finite. However, as is well-known, it is subtle and far from obvious whether the extra scalar mode actually decouples in the IR in the putative GR limit. A famous example is the van Dam-Veltman-Zakharov (vDVZ) discontinuity of massive gravity [10, 11] which can be attributed to the strongly coupled nature of a scalar mode, as elucidated in [12]. Massive gravity has three extra modes; a transverse spin one and a scalar mode. As it turned out, the scalar mode remains strongly coupled in the massless limit and thus the GR limit does not exist.

The original form of HL gravity has a similar strong coupling problem [9, 13]; there is an energy scale Λ_s above which the scalar mode self-coupling becomes strong. For HL gravity to flow to GR in the IR, the strong coupling scale Λ_s needs to

[6] As we have seen in Sect. 3.2.1, the scalar mode *does exist*, but it has $\omega = 0$. This can also be seen in the Stückelberg formalism, cf. (3.83). The scalar mode is neither exponentially growing nor oscillating to linear order. We will discuss more about the scalar mode in Sect. 3.2.4.

be sufficiently high so that the coupling becomes weak and virtually decouples in the IR. However, as it turned out, the naive estimate of the strong coupling scale yields a value too low, $\Lambda_s = \sqrt{\lambda_{\rm HL} - 1}\, M_P$, where $\lambda_{\rm HL}$ is an anisotropy parameter and the coefficient of the K^2 term; this is most manifest in the Stückelberg field formalism. The part of the Stückelberg field action, coming from the K^2 term, is proportional to $(\lambda_{\rm HL} - 1) M_P^2$. This is the most relevant part of the action in the IR. This implies that the coupling is $g = 1/(\sqrt{\lambda_{\rm HL} - 1}\, M_P)$ and thus irrelevant. Although the coupling becomes weak in the IR, the strong coupling scale Λ_s is too low, approaching zero in the putative GR limit $\lambda_{\rm HL} = 1$. In fact, as argued in [9], the situation appears to be even worse. The quadratic action in flat space lacks time derivatives, while there are time derivative interactions. This may imply that the high frequency modes are always strongly coupled. Indeed, one can refine the estimate of the strong coupling scale by making the space-time slightly curved (with the characteristic length scale L) which introduces a quadratic term with a time derivative. Then the refined strong coupling scale becomes anisotropic and is estimated to be $\Lambda_\omega = L^{-1/4} \Lambda_s^{3/4}$ and $\Lambda_p = L^{-3/4} \Lambda_s^{1/4}$ [9]. In the flat space limit $L \to \infty$, these are in fact zero.

In n-DBI, the situation turns out to be more subtle. The Stückelberg field action in flat space in unitary gauge yields

$$S_3 = \frac{1}{192\pi\lambda} \int d^4x \left[\frac{1}{2} (\Delta\dot{\varphi})^2 - (\dot{\varphi}\Delta\ddot{\varphi} - \partial_i\varphi\Delta\partial_i\dot{\varphi})\, \Delta\varphi + \frac{5}{12} \frac{(\Delta\dot{\varphi})^3}{\lambda M_P^2} \right], \quad (3.84)$$

to cubic order, as shown in Appendix B.3.3. The quadratic action is invariant under the scaling $E \to sE$, $t \to s^{-1}t$, $x \to s^{-1}x$, and $\varphi \to s^{-1}\varphi$. First note that the last interaction is irrelevant as compared to the middle ones. The higher order interactions have a similar structure. So the most "relevant" terms are of the middle type and actually (classically) marginal, with coupling $\lambda^{\frac{n-2}{2}}$, for the n-th order in φ.[7] However, due to quantum effects, these operators may become either relevant or irrelevant. For the cubic terms with coupling $g = \sqrt{\lambda}$, the beta function at 1-loop is of the form $\beta(g) = -cg^3$, from which one obtains[8]

$$g^2 = g_0^2 \left(1 + 2cg_0^2 \ln \frac{\Lambda}{M_P} \right)^{-1}, \quad (3.85)$$

where Λ is the energy scale and the UV cutoff scale is M_P. Therefore, the strong coupling scale is

$$\Lambda_s \sim M_P \exp\left(-\frac{1}{2c\lambda} \right). \quad (3.86)$$

This implies that in the putative GR limit $\lambda \to \infty$, the strong coupling scale is about the Planck scale and thus sufficiently high. However, for this coupling to be

---

[7] Most easily seen by redefining $\phi \to \sqrt{\lambda}\phi$.

[8] Here we assume that the IR divergences can be properly regularized.

(marginally) irrelevant, the constant c must be negative. Given the fact that this is a scalar field theory and the higher derivative nature of the operators, it seems plausible that this is actually the case. Therefore, n-DBI gravity is likely to be free from the strong coupling problem.

3.2.4 Discussion

The first main result is to have established that n-DBI gravity has three degrees of freedom: the usual two tensorial modes of the graviton plus a scalar mode which, however, is not a propagating particle with a definite dispersion relation. The second result is that the arguments for three pathological features of the original formulation of HL gravity do not carry over to n-DBI gravity; these arguments concern the absence of dynamics, a universal short distance instability, and a strong coupling issue. The reason why n-DBI looks healthier than HL gravity, in these respects, has a common ground: the nonlinear dependence on (spatial derivatives of) the lapse in the action.

The fact that the same arguments for the pathologies in HL gravity do not directly apply to n-DBI gravity does not mean, of course, that the latter is free of pathologies. Although the arguments presented herein are reassuring, in order to establish an healthy behavior, n-DBI gravity must be further studied. In particular, we found that the scalar mode grows linearly in time in the linear approximation and is thus at the threshold between stability (oscillation) and instability (exponential growth). In other words, the scalar mode can be either marginally stable or unstable.[9] The (in)stability depends on the details of the scalar mode interactions. We thus need to extend the analysis performed in Sect. 3.2.3.2 beyond linear order to establish the (in)stability of the scalar mode.

To study the effect of nonlinear interactions, we find it useful to work in the Einstein frame (1.135). Note that the background space-time remains flat under the frame change. A key observation is that, as suggested in Appendix B.4, the auxiliary field χ can essentially be thought of as the time derivative of the scalar mode, *i.e.*, $\chi \sim \dot{T} \sim \dot{\varphi}$. In other words, the proof of the stability amounts to showing the stability of the χ fluctuation. In fact, the equation of motion for χ in the $N^i = 0$ gauge

$$\partial_t(N^{-2}\dot{\chi}) + \Delta\chi - \frac{1}{3}\partial_t(N^{-1}K) - \frac{1}{12}V'(\chi) = 0 \tag{3.87}$$

---

[9] As an illustration, consider a simple mechanical model. For canonical kinetic term, a linear time growth is the behavior seen in a flat potential. The flat potential does not indicate an instability and is marginally stable. However, any potentials which are flat to quadratic order all lead to the linear time growth for perturbations in the linearized approximation. The simplest examples are cubic and (convex) quartic potentials. The former is clearly unstable (inverse-squared blow-up), whereas the latter is stable (oscillation).

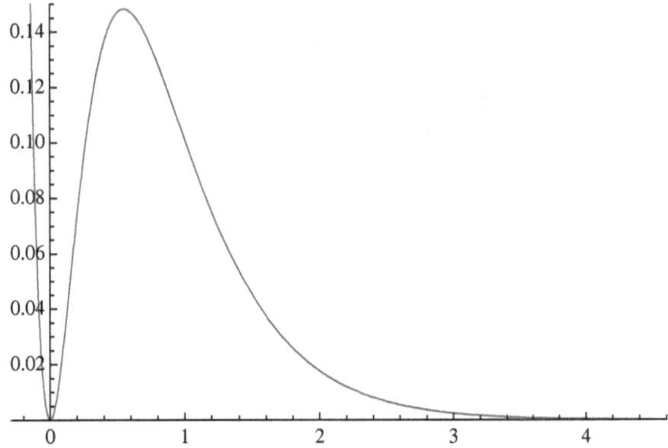

Fig. 3.1 The potential $V(\chi)$ defined in (3.2.3) with $q = 1$ (this figure is reproduced from [3])

seems to indicate the stability against small perturbations around $\chi = 0$; as is clear, the potential $V(\chi)$ plotted in Fig. 3.1 is very stable around $\chi = 0$. Note, however, that the lapse N and the spatial metric h_{ij} also furnish the scalar mode. Moreover, the auxiliary field χ is further constrained by the other equations of motion. So, strictly speaking, we need to deal with the coupled system of χ, N and h_{ij} (in the $N^i = 0$ gauge). There is, however, a shortcut: it would be sufficient to show that the solutions of (3.87) for χ in *generic backgrounds* are oscillatory.

The force in (3.87) consists of the strongly attractive force $\frac{1}{12}N^2V'(\chi)$, the repulsive force $-N^2\Delta\chi = N^2p^2\chi$, the friction $\partial_t \ln N^2\dot{\chi} \sim L^{-1}\dot{\chi}$, and the external force $\frac{1}{3}N^2\partial_t(N^{-1}K) \sim L^{-2}$. The last three forces may work against stability. These become more dominant when χ is smaller. However, the small χ behavior is well approximated by the linear perturbation. As shown in Appendix B.4, χ is marginally stable to linear order. Meanwhile, for larger χ, the attractive force, being exponential, may quickly become dominant, as χ increases, and pull it back to smaller χ. Hence it seems plausible that χ oscillates about $\chi = 0$. We therefore conclude that n-DBI gravity is likely to be stable against the scalar mode perturbations.

To close, we have considerably improved our understanding of the scalar mode and presented arguments for its healthy behavior. However, the scalar mode still remains somewhat elusive; it does not appear to behave as a conventional particle with a (non-)relativistic dispersion relation around near-flat space-times. So it is rather unclear what signatures exactly it might leave as an observable. On this score, however, we anticipate that the scalar mode could be the source of scalar perturbations in the Cosmic Microwave Background Radiation, when applied to inflationary cosmology [2], and we hope to return to this question in the near future.

References

1. C. Herdeiro, S. Hirano and Y. Sato, "n-DBI gravity", Phys. Rev. D 84, 124048 (2011) arXiv:1110.0832 [gr-qc].
2. Herdeiro, C., & Hirano, S. (2012). Scale invariance and a gravitational model with non-eternal inflation. *JCAP*, *1205*, 031. [arXiv:1109.1468] [hep-th].

3. Coelho, F. S., Herdeiro, C., Hirano, S., & Sato, Y. (2012). On the scalar graviton in n-DBI gravity. *Phys. Rev. D*, *86*, 064009. [arXiv:1205.6850] [hep-th].
4. T. Jacobson, "Einstein-aether gravity: A Status report", PoS QG -PH (2007) 020 arXiv:0801.1547 [gr-qc].
5. Horava, P. (2009). Quantum Gravity at a Lifshitz Point. *Phys. Rev.*, *D79*, 084008. [arXiv:0901.3775] [hep-th].
6. Stewart, J. M. (1990). Perturbations of Friedmann-Robertson-Walker cosmological models. *Class. Quant. Grav.*, *7*, 1169.
7. Blas, D., Pujolas, O., & Sibiryakov, S. (2010). Consistent Extension of Horava Gravity. *Phys. Rev. Lett.*, *104*, 181302. [arXiv:0909.3525] [hep-th].
8. Henneaux, M., Kleinschmidt, A., & Lucena, G. (2010). Gomez, "A dynamical inconsistency of Horava gravity". *Phys. Rev. D*, *81*, 064002. [arXiv:0912.0399] [hep-th].
9. Blas, D., Pujolas, O., & Sibiryakov, S. (2009). On the Extra Mode and Inconsistency of Horava Gravity. *JHEP*, *0910*, 029. [arXiv:0906.3046] [hep-th].
10. van Dam, H., & Veltman, M. J. G. (1970). Massive and massless Yang-Mills and gravitational fields. *Nucl. Phys. B*, *22*, 397.
11. V. I. Zakharov, "Linearized gravitation theory and the graviton mass", JETP Lett. 12, 312 (1970) [Pisma, Zh. Eksp. Teor. Fiz. 12, 447 (1970)].
12. Arkani-Hamed, N., Georgi, H., & Schwartz, M. D. (2003). Effective field theory for massive gravitons and gravity in theory space. *Annals Phys.*, *305*, 96. [hep-th/0210184].
13. Charmousis, C., Niz, G., Padilla, A., & Saffin, P. M. (2009). Strong coupling in Horava gravity. *JHEP*, *0908*, 070. arXiv:0905.2579 [hep-th].

Chapter 4
Summary

This thesis explores a role of space-time folation. One can consider the $(3 + 1)$-dimensional space-time as a 3-dimensional space parametrized by time; here, the 3-dimensional space (spatial hypersurface) is called *foliation* of space-time. Piling up foliations along a time direction, one can "reconstruct" space-time. Due to one of the guiding principles of general relativity, *general covariance* or *diffeomorphism invariance*, a shape of the space-time foliation is, in fact, not essential: it is a gauge artifact.

However, when considering a quantum notion of space-time based on the path-integral formalism, the space-time foliation seems to be a useful tool because it fits together well with causality. Causal Dynamical Triangulations (CDT) is a method to quantize space-time in this way: CDT allows us to sum all possible shapes of discretized space-time with the foliation structure in such a way as to import the concept of causality. The discretization of space-time is made in order to regularize short distance behaviors. In this thesis, we have considered, especially, the $(1 + 1)$-dimensional CDT and proposed an analytic technique to couple a dimer model. As it turns out, even when coupling dimers to the discretized space-time with the foliation structure, one can still find the critical point where the continuous space-time can be recovered, and at the critical point the dimers surve as matters coupled to space-time. Further, in the framework called *Generalized CDT* which is a generalization of CDT such that a creation and an annihilation of baby universes are allowed, we have suceeded in including extended interactions without ruining the causality structure based on a string field theory of CDT and found corresponding matrix-model descreptions as well.

On the other hands, it might be possible to consider that the space-time foliation is not an artifact but a physical "object" measured by some experiments. In such a case, the diffeomorphism invariance ought to be broken at the fundamental level. In this thesis, we have investigated this possibility as well based on n-DBI gravity. In n-DBI gravity, the existence of an everywhere time-like vector field n specifies a certain direction in space-time, which is an origin of breakdown of the diffeomorphism invariance. We have investigated black hole solutions of n-DBI gravity and found an

Y. Sato, *Space-Time Foliation in Quantum Gravity*, Springer Theses,
DOI: 10.1007/978-4-431-54947-5_4,

effect of the space-time foliation as a new integration constant appeared in blak hole solutions. Besides, we have nailed an extra scalar degree of freedom, originated with breaking diffeomorphism invariance, based on Dirac's theory of constrained systems. As a result, within our analysis this scalar mode does not show any pathological behaviors at a long distance scale.

At this stage it is hard to tell if the space-time foliation is merely an artifact introduced for convenience or a genuinly measurable "object", but in any case it seems to the author that the space-time foliation might be a key idea when understanding a quantum notion of space-time. This is because through the foliation one can take in an important property quantum gravity is supposed to posess, causality, in a natural manner; as far as we have explored in this thesis, we could not find any pathologies related to it, although our argument is still model dependent. Anyway, in order to reach at the truth of nature, what one can do is to investigate logical possibilities, consider deeply as long as possible and compare them with a bunch of experiments. Even though each step might only have a tiny progress and most of the cases might not bear any fruit, as a result of thousands of random walks gradually human being is approaching the correct way; in the end of the day, our understandings coverge to one single picture of the real space-time structure. Only thing we are in no doubt at the current stage is that the results shown in this thesis is a part of such tremendusly many random walks.

Appendix A
Notation and Formulae

In this Appendix we list our notion and formulae used in this thesis. Since we work on $1+1$ and $3+1$ dimensions, here we generalize the space-time dimensionality to $d+1$ dimensions, and we use the signature $(-,+,+,\cdots,+)$ for a Lorentzian metric $g_{\mu\nu}$. Greek indices run from 0 to d in the $d+1$-dimensional space-time, and we also use Latin indices for the d-dimensional Euclidean space. Indices of all $d+1$-dimensional tensors, pseudo-tensors and densities are lowered and raised by the metric $g_{\mu\nu}$ and its inverse $g^{\mu\nu}$, respectively. In addition, we apply the Einstein summation convention. The covariant derivative compatible with $g_{\mu\nu}$ is denoted by ∇_λ, and we use Δ as the Laplacian, $g^{\mu\nu}\nabla_\mu\nabla_\nu$. We write the cosmological constant and Newton constant as Λ and G_{d+1}, respectively. Then here we go.

Christoffel symbol:

$$\Gamma^\lambda_{\mu\nu} = \frac{1}{2} g^{\lambda\sigma} \left(\partial_\mu g_{\nu\sigma} + \partial_\nu g_{\mu\sigma} - \partial_\sigma g_{\mu\nu} \right). \tag{A.1}$$

Covariant derivative acting on covariant and contravariant vectors:

$$\nabla_\mu t^\lambda = \partial_\mu t^\lambda + \Gamma^\lambda_{\mu\nu} t^\nu, \quad \nabla_\mu w_\nu = \partial_\mu w_\nu - \Gamma^\lambda_{\mu\nu} w_\lambda. \tag{A.2}$$

Covariant derivative acting on density:

$$\nabla_\sigma p^{\mu\nu} = \partial_\sigma p^{\mu\nu} + \Gamma^\mu_{\sigma\tau} p^{\tau\nu} + \Gamma^\nu_{\sigma\tau} p^{\mu\tau} - \Gamma^\tau_{\tau\sigma} p^{\mu\nu}. \tag{A.3}$$

Riemann tensor:

$$R^\lambda{}_{\mu\nu\sigma} = \partial_\nu \Gamma^\lambda_{\mu\sigma} - \partial_\sigma \Gamma^\lambda_{\mu\nu} + \Gamma^\eta_{\mu\sigma} \Gamma^\lambda_{\nu\eta} - \Gamma^\eta_{\mu\nu} \Gamma^\lambda_{\sigma\eta}. \tag{A.4}$$

Ricci tensor and Ricci scalar:

$$R_{\mu\nu} = R^\lambda{}_{\mu\lambda\nu}, \quad R = g^{\mu\nu} R_{\mu\nu}. \tag{A.5}$$

Y. Sato, *Space-Time Foliation in Quantum Gravity*, Springer Theses,
DOI: 10.1007/978-4-431-54947-5,

Commutators of covariant derivatives:

$$[\nabla_\mu, \nabla_\nu] t^\lambda = R^\lambda{}_{\sigma\mu\nu} t^\sigma, \tag{A.6}$$

$$[\nabla_\mu, \nabla_\nu] w_\lambda = -R^\sigma{}_{\lambda\mu\nu} w_\sigma. \tag{A.7}$$

Variation of metric:

$$\delta g^{\mu\nu} = -g^{\mu\sigma} g^{\nu\tau} \delta g_{\sigma\tau}. \tag{A.8}$$

Variation of Christoffel symbol:

$$\delta \Gamma^\lambda_{\mu\nu} = \frac{1}{2} g^{\lambda\sigma} \left(\nabla_\mu \delta g_{\nu\sigma} + \nabla_\nu \delta g_{\mu\sigma} - \nabla_\sigma \delta g_{\mu\nu} \right). \tag{A.9}$$

Variation of Ricci tensor:

$$\delta R_{\mu\nu} = \nabla_\lambda \delta \Gamma^\lambda_{\mu\nu} - \nabla_\nu \delta \Gamma^\lambda_{\lambda\mu}. \tag{A.10}$$

Variation of Ricci scalar:

$$\delta R = -R^{\mu\nu} \delta g_{\mu\nu} + (\nabla^\mu \nabla^\nu - g^{\mu\nu} \Delta) \delta g_{\mu\nu}. \tag{A.11}$$

Variation of Laplacian acting on scalar:

$$\delta(\Delta e) = \Delta(\delta e) - \nabla^\mu \nabla^\nu e \delta g_{\mu\nu} - \nabla^\mu e \nabla^\nu \delta g_{\mu\nu} + \frac{1}{2} g^{\mu\nu} \nabla^\tau e \nabla_\tau \delta g_{\mu\nu}. \tag{A.12}$$

Einstein-Hilbert action with Gibbons-Hawking-York boundary term:

$$S = -\frac{1}{16\pi G_{d+1}} \int_{\mathcal{M}} d^{d+1}x \sqrt{-g} \, (R - 2\Lambda + \mathcal{K}) \tag{A.13}$$

$$= -\frac{1}{16\pi G_{d+1}} \int_{\mathcal{M}} d^{d+1}x \sqrt{-g} \, (R - 2\Lambda) + \frac{1}{8\pi G_{d+1}} \int_{\partial\mathcal{M}} d^d x \sqrt{h} K. \tag{A.14}$$

Appendix B
Computations in *n*-DBI Gravity

B.1 Derivation of the Solutions

Taking the ansatz

$$ds^2 = -N^2(r)dt^2 + e^{2f(r)}\left(dr + e^{2g(r)}dt\right)^2 + r^2 d\Omega_2, \tag{B.1}$$

it follows that ($' \equiv d/dr$):

$$\begin{aligned} K_{ij} &= -\frac{e^{2g}}{N}\,\mathrm{diag}\left\{e^{2f}(f+2g)', r, r\sin^2\theta\right\}, \\ K &= -\frac{e^{2g}}{N}\left(\frac{2}{r} + (f+2g)'\right), \end{aligned} \tag{B.2}$$

$$R_{ij} = \mathrm{diag}\left\{\frac{2f'}{r}, \frac{rf' + e^{2f} - 1}{e^{2f}}, \sin^2\theta\frac{rf' + e^{2f} - 1}{e^{2f}}\right\}, \tag{B.3}$$

$$R = \frac{2e^{-2f}}{r^2}\left(2rf' + e^{2f} - 1\right), \tag{B.4}$$

Thus,

$$\mathscr{R} = \frac{2}{r^2}\left[1 - \left(re^{-2f}\right)' - \frac{\left(re^{4g}\right)'}{N^2} - \frac{2rf'e^{4g}}{N^2} - \frac{e^{-f}}{N}\left(r^2N'e^{-f}\right)'\right]. \tag{B.5}$$

From the effective Lagrangian (3.16), the A equation of motion can be solved straightaway to yield

$$A' = Q\frac{Ne^f}{r^2}, \tag{B.6}$$

Y. Sato, *Space-Time Foliation in Quantum Gravity*, Springer Theses,
DOI: 10.1007/978-4-431-54947-5, © Springer Japan 2014

where Q is an integration constant. The g equation of motion yields the compact relation

$$\left(\frac{1}{\sqrt{1+\frac{G_4}{6\lambda}\mathscr{R}}}\right)' = \frac{(f+\ln N)'}{\sqrt{1+\frac{G_4}{6\lambda}\mathscr{R}}}, \tag{B.7}$$

and can be integrated to yield

$$\left(1+\frac{G_4}{6\lambda}\mathscr{R}\right)^{-1/2} = \frac{Ne^f}{C}, \tag{B.8}$$

where C is an integration constant. More explicitly, this equation may be written as

$$\left(re^{4g+2f}\right)' = N^2e^{2f}\left(1-(re^{-2f})'\right) - Ne^f\left(r^2e^{-f}N'\right)' + \frac{3\lambda}{G_4}r^2\left(N^2e^{2f}-C^2\right). \tag{B.9}$$

The f and N equations of motion, upon using (B.6) and (B.8), read, respectively

$$\left(re^{4g+2f}\right)' = re^{-2f}\left(e^{2f}N^2\right)' - \frac{Ne^f}{2}\left(r^2e^{-f}N'\right)' + \frac{r^2e^{-f}N'}{2}(Ne^f)' - \frac{3\lambda}{G_4}r^2\left(C^2-CqNe^f\right) + \frac{G_4CQ^2Ne^f}{2r^2}, \tag{B.10}$$

$$\left(re^{4g+2f}\right)' = -\frac{1}{4}\left(r^2\left(e^{-2f}\right)'N^2e^{2f}\right)' - \frac{3\lambda}{G_4}r^2\left(C^2-CqNe^f\right) + \frac{G_4CQ^2Ne^f}{2r^2}. \tag{B.11}$$

Equation (B.11) is the Hamiltonian constraint (3.3), after using (B.8) and (B.9).

B.1.1 Solutions with Constant $\mathscr{R}$

To proceed we take the combination $Ne^f = \tilde{C} =$ constant which implies that $\mathscr{R}$ is constant. We shall address the general solution in the next subsection, but it turns out that the most interesting solution are found in this subset. With this choice, we observe from Eq. (B.8) that $\mathscr{R} =$ constant. From the resulting equations of motion, equating (B.9) with either (B.10) or (B.11) (which become identical), we find the ODE:

$$Y'' + \frac{6}{r}Y' + \frac{4}{r^2}Y = \frac{12\lambda}{G_4}\left(1-\frac{qC}{\tilde{C}}\right) - \frac{2G_4CQ^2}{\tilde{C}r^4}, \tag{B.12}$$

where $Y \equiv e^{-2f} - 1$. It is now straightforward to obtain the exact solution. It reads (3.17), where $\tilde{C}$ has been eliminated by rescaling C and the time coordinate.

B.1.2 Generic Solution

Since the set of equations we are solving is a second order ODE with three unknowns, we expect a total of six integration constants. The constant $\mathscr{R}$ solution exhibited below has only five integration constants and thus it is not the most general one. The latter can be obtained observing that the Eqs. (B.10) and (B.11) imply

$$-\left(r^{-2}(\log N)'\right)' = \left(r^{-2}f'\right)' \qquad \Longrightarrow \qquad Ne^{f} = \tilde{C}e^{\frac{1}{3}C_4 r^3}. \tag{B.13}$$

C_4 is the sixth integration constant, which was absent in the constant $\mathscr{R}$ solution. Similarly to the constant $\mathscr{R}$ case, it is straightforward to find a second order ODE:

$$W'' + \frac{6}{r}W' + \frac{4}{r^2}W = \frac{4e^{\frac{2C_4r^3}{3}}}{r^2} + \frac{12\lambda}{G_4}\left(e^{\frac{2C_4r^3}{3}} - \frac{qCe^{\frac{C_4r^3}{3}}}{\tilde{C}}\right) - \frac{2G_4CQ^2e^{\frac{C_4r^3}{3}}}{\tilde{C}r^4}, \tag{B.14}$$

where we defined $W \equiv e^{-2\left(f-\frac{C_4r^3}{3}\right)}$. This can be integrated to give explicit solutions. As they are not very illuminating, however, we will not present them here. Indeed, the solutions with $C_4 \neq 0$ seem rather exotic, since (B.8) and (B.13) imply that their asymptotic behavior at $r = +\infty$ is very different from that of Einstein gravity: the $C_4 < 0$ solutions have a curvature singularity at $r = +\infty$ and thus we regard these solutions as unphysical; the $C_4 > 0$ solutions have the maximal negative curvature $\mathscr{R} = -6\lambda/G_4$ at $r = +\infty$. Although they are interesting in their own right, we shall not discuss these solutions further herein.

B.2 Computations of Constraint Algebra

In this Appendix, we give an explicit computation of the constraint algebra and classification of the constraints presented in Sect. 3.2.1. To facilitate the computation, we introduce smooth test vector fields, $\xi^\mu = (\xi^0, \xi^i)$ and $\eta^\mu = (\eta^0, \eta^i)$, which fall off fast enough to suppress all the boundary contributions [1]. Henceforth, we define the smeared constraints:

$$\hat{\Phi}_1(\xi^0) = \int d^3x\,\xi^0(x)\Phi_1(x), \tag{B.15}$$

$$\hat{\Phi}_2(\xi^i) = \int d^3x\,\xi^i(x)\Phi_{2i}(x), \tag{B.16}$$

$$\hat{\Phi}_3(\xi^0) = \int d^3x\,\xi^0(x)\Phi_3(x), \tag{B.17}$$

$$\hat{\Phi}_4(\xi^0) = \int d^3x\,\xi^0(x)\Phi_4(x), \tag{B.18}$$

$$\hat{\Phi}_5(\xi^i) = \int d^3x\,\xi^i(x)\Phi_{5i}(x), \tag{B.19}$$

$$\hat{\Phi}_6(\xi^0) = \int d^3x\,\xi^0(x)\Phi_6(x), \tag{B.20}$$

$$\hat{\Phi}_N^{(G)}(\xi^i) = \int d^3x \xi^i(x) \Phi_{Ni}^{(G)}(x) = \int d^3x \xi^i(x)(-\Phi_1 \partial_i N)(x), \tag{B.21}$$

$$\hat{\Phi}_e^{(G)}(\xi^i) = \int d^3x \xi^i(x) \Phi_{ei}^{(G)}(x) = \int d^3x \xi^i(x)(-\Phi_3 \partial_i e)(x), \tag{B.22}$$

$$\hat{\Phi}_{\mathbf{N}}^{(G)}(\xi^i) = \int d^3x \xi^i(x) \Phi_{\mathbf{N}i}^{(G)}(x) = \int d^3x \xi^i(x)\Big[-\left(\Phi_2^i \nabla_i N^j + \nabla_j \left(\Phi_{2i} N^j\right)\right)(x)\Big], \tag{B.23}$$

$$\hat{\Phi}_5'(\xi^i) = \hat{\Phi}_5(\xi^i) + \hat{\Phi}_N^{(G)}(\xi^i) + \hat{\Phi}_e^{(G)}(\xi^i) = \int d^3x \xi^i(x) \tilde{\Phi}_{5i}(x), \tag{B.24}$$

where Φ_1, Φ_{2i}, Φ_3, Φ_4, Φ_{5i}, Φ_6 and $\tilde{\Phi}_{5i}$ are the constraints defined in Sect. 3.2.1. The idea is to compute the commutators like

$$\left\{\hat{\Phi}_4(\xi^0), \hat{\Phi}_5(\eta^j)\right\} = \int d^3y d^3x \xi^0(y) \eta^j(x) \left\{\Phi_4(y), \Phi_{5j}(x)\right\}, \tag{B.25}$$

and read off the algebra of local constraints from the R.H.S. The basic non-vanishing Poisson brackets are given by

$$\left\{p^{ij}(y), h_{kl}(x)\right\} = \frac{1}{2}\left(\delta_k^i \delta_l^j + \delta_l^i \delta_k^j\right) \delta(y-x), \tag{B.26}$$

$$\{p_N(y), N(x)\} = \delta(y-x), \tag{B.27}$$

$$\left\{p_{\mathbf{N}}^i(y), N_j(x)\right\} = \delta_j^i \delta(y-x), \tag{B.28}$$

$$\{p_e(y), e(x)\} = \delta(y-x). \tag{B.29}$$

To compute the Poisson brackets of constraints, we take the variations of the smeared constraints:

$$\frac{\delta \hat{\Phi}_1(\xi^0)}{\delta p_N} = \xi^0, \tag{B.30}$$

$$\frac{\delta \hat{\Phi}_2(\xi^i)}{\delta p_{\mathbf{N}i}} = \xi^i, \tag{B.31}$$

$$\frac{\delta \hat{\Phi}_3(\xi^0)}{\delta p_e} = \xi^0, \tag{B.32}$$

$$\begin{aligned}\frac{\delta \hat{\Phi}_4(\xi^0)}{\delta h_{mn}} &= \xi^0 \Big[\frac{1}{2} h^{mn} \Phi_4 - \frac{\kappa}{e\sqrt{h}} h^{mn} \left(p^{ab} p_{ab} - \frac{1}{2} p^2\right) \\ &\quad + \frac{2\kappa}{e\sqrt{h}} \left(p^{ml} p_l^n - \frac{1}{2} p^{mn} p\right) \\ &\quad + \frac{e}{\kappa} \sqrt{h} R^{mn} - \frac{\sqrt{h}}{\kappa} (\nabla^m \nabla^n e)\Big] \\ &\quad + \frac{\sqrt{h}}{\kappa}\Big[-(\nabla^m \nabla^n \xi^0) e + h^{mn} (\nabla^a e)(\nabla_a \xi^0) + h^{mn} e(\Delta \xi^0)\Big],\end{aligned} \tag{B.33}$$

$$\frac{\delta \hat{\Phi}_4(\xi^0)}{\delta p^{mn}} = \xi^0 \Big[\frac{2\kappa}{e\sqrt{h}} \left(p_{mn} - \frac{1}{2} h_{mn} p\right)\Big], \tag{B.34}$$

$$\frac{\delta \hat{\Phi}_4(\xi^0)}{\delta e} = -\xi^0 \left[\frac{\kappa}{e^2\sqrt{h}} \left(p^{ab} p_{ab} - \frac{1}{2} p^2 \right) + \frac{\sqrt{h}}{\kappa} R + \frac{6\lambda\sqrt{h}}{\kappa G_4} \left(1 - \frac{1}{e^2} \right) \right] + \frac{2\sqrt{h}}{\kappa} \Delta \xi^0, \tag{B.35}$$

$$\frac{\delta \hat{\Phi}_5(\xi^i)}{\delta h_{mn}} = -(\nabla_l \xi^i) p^{lj} (\delta^m_i \delta^n_j + \delta^m_j \delta^n_i) + \nabla_l (p^{mn} \xi^l) = \pounds_\xi p^{mn}, \tag{B.36}$$

$$\frac{\delta \hat{\Phi}_5(\xi^i)}{\delta p^{mn}} = -h_{li} (\nabla_j \xi^l)(\delta^i_m \delta^j_n + \delta^i_n \delta^j_m) = -\pounds_\xi h_{mn}, \tag{B.37}$$

$$\begin{aligned} \frac{\delta \hat{\Phi}_6(\xi^0)}{\delta h_{mn}} = \xi^0 \Bigg[& \frac{1}{2} h^{mn} \Phi_6 + \frac{N\kappa}{e^2\sqrt{h}} h^{mn} \left(p^{ab} p_{ab} - \frac{1}{2} p^2 \right) \\ & - \frac{2\kappa N}{e^2\sqrt{h}} \left(p^{ml} p^n_l - \frac{1}{2} p^{mn} p \right) \\ & + \frac{N\sqrt{h}}{\kappa} R^{mn} - \frac{\sqrt{h}}{\kappa} (\nabla^m \nabla^n N) \Bigg] \\ & + \frac{\sqrt{h}}{\kappa} \left[-N(\nabla^m \nabla^n - h^{mn} \Delta)\xi^0 + h^{mn} (\nabla^l N)(\nabla_l \xi^0) \right], \end{aligned} \tag{B.38}$$

$$\frac{\delta \hat{\Phi}_6(\xi^0)}{\delta p^{mn}} = -\xi^0 \left[\frac{2N\kappa}{e^2\sqrt{h}} \left(p_{mn} - \frac{1}{2} h_{mn} p \right) \right], \tag{B.39}$$

$$\frac{\delta \hat{\Phi}_6(\xi^0)}{\delta N} = -\xi^0 \left[\frac{\kappa}{e^2\sqrt{h}} \left(p^{ab} p_{ab} - \frac{1}{2} p^2 \right) + \frac{\sqrt{h}}{\kappa} R + \frac{6\lambda\sqrt{h}}{\kappa G_4} \left(1 - \frac{1}{e^2} \right) \right] + 2\frac{\sqrt{h}}{\kappa} \Delta(\xi^0), \tag{B.40}$$

$$\frac{\delta \hat{\Phi}_6(\xi^0)}{\delta e} = \xi^0 \left[\frac{2N\kappa}{e^3\sqrt{h}} \left(p^{ab} p_{ab} - \frac{1}{2} p^2 \right) - \frac{12\lambda N\sqrt{h}}{\kappa e^3 G_4} \right], \tag{B.41}$$

$$\frac{\delta \hat{\Phi}^{(G)}_N(\xi^i)}{\delta N} = \partial_i (\xi^i p_N), \tag{B.42}$$

$$\frac{\delta \hat{\Phi}^{(G)}_N(\xi^i)}{\delta p_N} = -\xi^i \partial_i N = -\pounds_\xi N, \tag{B.43}$$

$$\frac{\delta \hat{\Phi}^{(G)}_e(\xi^i)}{\delta e} = \partial_i (\xi^i p_e), \tag{B.44}$$

$$\frac{\delta \hat{\Phi}^{(G)}_e(\xi^i)}{\delta p_e} = -\xi^i \partial_i e = -\pounds_\xi e, \tag{B.45}$$

$$\frac{\delta \hat{\Phi}^{(G)}_{\mathbf{N}}(\xi^i)}{\delta N^j} = \nabla_i (\xi^i p_{\mathbf{N}j}) + (\nabla_j \xi^i) p_{\mathbf{N}i}, \tag{B.46}$$

$$\frac{\delta \hat{\Phi}^{(G)}_{\mathbf{N}}(\xi^i)}{\delta p_{\mathbf{N}}} = -\xi^i \nabla_i N^j + N^i \nabla_i \xi^j = -\pounds_\xi N^j. \tag{B.47}$$

Using these, it is tedious but straightforward to compute the constraint algebra:

$$\{\Phi_1(y), \Phi_6(x)\} = -\left[\frac{\kappa}{e^2\sqrt{h}}\left(p^{ab}p_{ab} - \frac{1}{2}p^2\right) + \frac{\sqrt{h}}{\kappa}R + \frac{6\lambda\sqrt{h}}{\kappa G_4}\left(1 - \frac{1}{e^2}\right)\right] \delta(y-x) + 2\frac{\sqrt{h}}{\kappa}\Delta\delta(y-x), \tag{B.48}$$

$$\left\{\Phi_1(y), \Phi^{(G)}_{Nj}(x)\right\} = \Phi_1(x)\partial_{y^j}\delta(y-x), \tag{B.49}$$

$$\{\Phi_3(y), \Phi_4(x)\} = -\left[\frac{\kappa}{e^2\sqrt{h}}\left(p^{ab}p_{ab} - \frac{1}{2}p^2\right) + \frac{\sqrt{h}}{\kappa}R + \frac{6\lambda\sqrt{h}}{\kappa G_4}\left(1 - \frac{1}{e^2}\right)\right] \delta(y-x) + \frac{2\sqrt{h}}{\kappa}\Delta\delta(y-x), \tag{B.50}$$

$$\{\Phi_3(y), \Phi_6(x)\} = \left[\frac{2N\kappa}{e^3\sqrt{h}}\left(p^{ab}p_{ab} - \frac{1}{2}p^2\right) - \frac{12\lambda N\sqrt{h}}{\kappa e^3 G_4}\right]\delta(y-x), \tag{B.51}$$

$$\left\{\Phi_3(y), \Phi^{(G)}_{ej}(x)\right\} = \Phi_3(x)\partial_{y^j}\delta(y-x), \tag{B.52}$$

$$\{\Phi_4(y), \Phi_4(x)\} = \Phi_5^k(y)\partial_{y^k}\delta(y-x) - \Phi_5^k(x)\partial_{x^k}\delta(y-x) - \frac{p}{e}\partial_{y^k}e\partial_{y_k}\delta(y-x) + \frac{p}{e}\partial_{x^k}e\partial_{x_k}\delta(y-x), \tag{B.53}$$

$$\left\{\Phi_4(y), \Phi_{5j}(x)\right\} = \Phi_4(x)\partial_{y^j}\delta(y-x) - \frac{\partial_{x^j}e}{N}\Phi_6(x)\delta(y-x) - 2\frac{\sqrt{h}}{\kappa}\partial_{x^j}e\left(\Delta - \frac{\Delta N}{N}\right)\delta(y-x), \tag{B.54}$$

$$\begin{aligned}\{\Phi_4(y), \Phi_6(x)\}\} = {}& \frac{2Np_{mn}}{e}\left[2R^{mn} - \left(\frac{\nabla^m\nabla^m e}{e} + \frac{\nabla^m\nabla^n N}{N}\right)\right]\delta(y-x) \\ & - \frac{Np}{e}\left[R - \frac{6\lambda}{G_4}\left(1 - \frac{q}{e}\right)\right]\delta(y-x) \\ & - \frac{2Np_{mn}(y)}{e}\nabla^m_{(y)}\partial_{y_n}\delta(y-x) - \frac{2Np_{mn}(x)}{e}\nabla^m_{(x)}\partial_{x_n}\delta(y-x) \\ & - \left(\frac{Np}{e}\right)\frac{\partial_{y^m}N}{N}\partial_{y_m}\delta(y-x) - \left(\frac{Np}{e}\right)\frac{\partial_{x^m}e}{e}\partial_{x_m}\delta(y-x),\end{aligned} \tag{B.55}$$

$$\left\{\Phi_4(y), \Phi^{(G)}_{ej}(x)\right\} = \frac{\partial_{x^j}e}{N}\Phi_6(x)\delta(y-x) + 2\frac{\sqrt{h}}{\kappa}\partial_{x^j}e\left(\Delta - \frac{\Delta N}{N}\right)\delta(y-x), \tag{B.56}$$

$$\left\{\Phi_{5j}(y), \Phi_{5i}(x)\right\} = \Phi_{5j}(x)\partial_{y^i}\delta(y-x) - \Phi_{5i}(y)\partial_{x^j}\delta(y-x), \tag{B.57}$$

$$\left\{\Phi_6(y), \Phi_{5j}(x)\right\} = \Phi_6(x)\partial_{y^j}\delta(y-x) - \frac{\partial_{x^j}N}{N}\Phi_6(x)\delta(y-x) - 2\frac{\sqrt{h}}{\kappa}\partial_{x^j}N\left(\Delta - \frac{\Delta N}{N}\right)\delta(y-x) - \frac{2N\sqrt{h}}{\kappa}\partial_{y^j}e\left(\frac{B(h,p)}{e^3}\right)\delta(y-x), \tag{B.58}$$

$$\begin{aligned}\{\Phi_6(y), \Phi_6(x)\} = &-\partial_{y_m}\left(\frac{2N^2}{e^2}\right)p_{mn}(y)\partial_{y_n}\delta(y-x) + \partial_{x_m}\left(\frac{2N^2}{e^2}\right) \\ &p_{mn}(x)\partial_{x_n}\delta(y-x) \\ &-\left(\frac{N}{e}\right)^2\Phi_{5n}(y)\partial_{y_n}\delta(y-x) + \left(\frac{N}{e}\right)^2\Phi_{5n}(x)\partial_{x_n}\delta(y-x) \\ &+\frac{Np\partial_{y^m}N}{e^2}\partial_{y_m}\delta(y-x) - \frac{Np\partial_{x^m}N}{e^2}\partial_{x_m}\delta(y-x), \end{aligned} \tag{B.59}$$

$$\left\{\Phi_6(y), \Phi_{Nj}^{(G)}(x)\right\} = \frac{\partial_{x^j}N}{N}\Phi_6(x)\delta(y-x) + 2\frac{\sqrt{h}}{\kappa}\partial_{x^j}N\left(\Delta - \frac{\Delta N}{N}\right)\delta(y-x), \tag{B.60}$$

$$\left\{\Phi_6(y), \Phi_{ej}^{(G)}(x)\right\} = \frac{2N\sqrt{h}}{\kappa}\partial_{y^j}e\left(\frac{B(h,p)}{e^3}\right)\delta(y-x), \tag{B.61}$$

$$\left\{\Phi_{Nj}^{(G)}(y), \Phi_{Ni}^{(G)}(x)\right\} = \Phi_{Nj}^{(G)}(x)\partial_{y^i}\delta(y-x) - \Phi_{Ni}^{(G)}(y)\partial_{x^j}\delta(y-x), \tag{B.62}$$

$$\left\{\Phi_{ej}^{(G)}(y), \Phi_{ei}^{(G)}(x)\right\} = \Phi_{ej}^{(G)}(x)\partial_{y^i}\delta(y-x) - \Phi_{ei}^{(G)}(y)\partial_{x^j}\delta(y-x), \tag{B.63}$$

$$\begin{aligned}\{\Phi_1(y), \Phi_{2j}(x)\} &= \{\Phi_{2i}(y), \Phi_{2j}(x)\} = \{\Phi_3(y), \Phi_{2j}(x)\} = \{\Phi_4(y), \Phi_{2j}(x)\} \\ &= \{\Phi_{5i}(y), \Phi_{2j}(x)\} = \{\Phi_6(y), \Phi_{2j}(x)\} = \left\{\Phi_{Ni}^{(G)}(y), \Phi_{2j}(x)\right\} \\ &= \left\{\Phi_{ei}^{(G)}(y), \Phi_{2j}(x)\right\} = 0. \end{aligned} \tag{B.64}$$

In order to classify the class of the constraints, it is more appropriate to choose

$$(\Phi_1, \Phi_{2j}, \Phi_3, \Phi_4, \tilde{\Phi}_{5j}, \Phi_6) \tag{B.65}$$

as a set of independent constraints. $\tilde{\Phi}_{5j}$ is defined in (B.24) and given by a linear combination $\Phi_{5i} - \partial_i N\Phi_1 - \partial_i e\Phi_3$. As is clear from the above computation, Φ_{2j} and $\tilde{\Phi}_{5j}$ commute with all the constraints:

$$\begin{aligned}\{\Phi_1(y), \Phi_{2j}(x)\} &= \{\Phi_{2i}(y), \Phi_{2j}(x)\} = \{\Phi_3(y), \Phi_{2j}(x)\} = \{\Phi_4(y), \Phi_{2j}(x)\} \\ &= \left\{\tilde{\Phi}_{5i}(y), \Phi_{2j}(x)\right\} = \{\Phi_6(y), \Phi_{2j}(x)\} = 0. \end{aligned} \tag{B.66}$$

$$\left\{\Phi_1(y), \tilde{\Phi}_{5j}(x)\right\} = \Phi_1(x)\partial_{y^j}\delta(y-x) \approx 0, \tag{B.67}$$

$$\left\{\Phi_3(y), \tilde{\Phi}_{5j}(x)\right\} = \Phi_3(x)\partial_{y^j}\delta(y-x) \approx 0, \tag{B.68}$$

$$\left\{\Phi_4(y), \tilde{\Phi}_{5j}(x)\right\} = \Phi_4(x)\partial_{y^j}\delta(y-x) \approx 0, \tag{B.69}$$

$$\left\{\tilde{\Phi}_{5i}(y), \tilde{\Phi}_{5j}(x)\right\} = \tilde{\Phi}_{5i}(x)\partial_{y^j}\delta(y-x) - \tilde{\Phi}_{5j}(y)\partial_{x^i}\delta(y-x) \approx 0, \tag{B.70}$$

$$\left\{\Phi_6(y), \tilde{\Phi}_{5j}(x)\right\} = \Phi_6(x)\partial_{y^j}\delta(y-x) \approx 0. \tag{B.71}$$

It is easy to show that the set (B.65) is complete. Namely, the time flows of the secondary constraints do not give rise to any new constraints:

$$\dot{\Phi}_4(x) = \int d^3y \left\{\mathscr{H}^{e(0)}_{\mathrm{nDBI}}(y), \Phi_4(x)\right\} + \sum_{a=1,3} \int d^3y \left\{\Phi_a(y), \Phi_4(x)\right\} \lambda_a$$
$$+ \int d^3y \left\{\Phi_2^i(y), \Phi_4(x)\right\} \lambda_{2i} \approx 0, \tag{B.72}$$

$$\dot{\tilde{\Phi}}_{5j}(x) = \int d^3y \left\{\mathscr{H}^{e(0)}_{\mathrm{nDBI}}(y), \tilde{\Phi}_{5j}(x)\right\} + \sum_{a=1,3} \int d^3y \left\{\Phi_a(y), \tilde{\Phi}_{5j}(x)\right\} \lambda_a$$
$$+ \int d^3y \left\{\Phi_2^i(y), \tilde{\Phi}_{5j}(x)\right\} \lambda_{2i} \approx 0, \tag{B.73}$$

$$\dot{\Phi}_6(x) = \int d^3y \left\{\mathscr{H}^{e(0)}_{\mathrm{nDBI}}(y), \Phi_6(x)\right\} + \sum_{a=1,3} \int d^3y \left\{\Phi_a(y), \Phi_6(x)\right\} \lambda_a$$
$$+ \int d^3y \left\{\Phi_2^i(y), \Phi_6(x)\right\} \lambda_{2i} \approx 0. \tag{B.74}$$

Since the Hamiltonian density takes the form

$$\mathscr{H}^{e(0)}_{\mathrm{nDBI}} = -\left(N\Phi_4 + N^j\Phi_{5j}\right) - \frac{2}{\kappa}\sqrt{h}\,(e\Delta N - N\Delta e)\,, \tag{B.75}$$

one can see that (B.72) and (B.74) determine λ_1 and λ_3, while (B.73) is trivially satisfied. Hence, there are no additional constraints from these time flows. Having established the completeness of the set (B.65), we conclude that Φ_{2j} and $\tilde{\Phi}_{5j}$ are first class, and the rest are second class.

Finally, let us end this Appendix with a comment on the generator of the spatial diffeomorphism. The generator $\mathscr{G}(\xi^i)$ of the spatial diffeomorphism acts on a phase-space variable A as

$$\left\{A(y), \mathscr{G}(\xi^i)\right\} = \pounds_\xi A(y). \tag{B.76}$$

Since p_N, $p^i_{\mathbf{N}}$, and p_e are primary constraints, we only need to consider the reduced set of phase-space variables, $(h_{ij}, p^{ij}, N, \mathbf{N}, e)$. The spatial diffeomorphisms for this set are generated by $(\Phi_5, -\Phi_5, \Phi_N^{(G)}, \Phi_{\mathbf{N}}^{(G)}, \Phi_e^{(G)})$. They are indeed all generated by the first class constraints, as can be seen from (B.21)–(B.24).

B.3 Computational Details of Perturbations

In this Appendix, we show some details of the computations in the perturbative analysis of the scalar mode in Sects. 3.2.1 and 3.2.2.

B.3.1 Perturbation of the Equations of Motion

The linearised version of Eqs. (3.72)–(3.75) and (3.76) can be obtained in a fashion similar to [2]. To the approximation explained in Sect. 3.2.3.2, we find, in the gauge $N^i = 0$,

$$\dot{\gamma}_{ij} - 2\bar{N}\kappa_{ij} - 2\bar{K}_{ij}n = 0, \tag{B.77}$$

$$2\bar{K}_{ij}\kappa^{ij} - 2\kappa\bar{K} - \nabla^i\nabla^j\gamma_{ij} + \Delta\gamma = \frac{G_4}{6\lambda}\Delta\alpha, \tag{B.78}$$

$$\nabla^j\kappa_{ij} - \nabla_i\kappa - \frac{3}{2}\bar{K}^{jk}\nabla_i\gamma_{jk} + \bar{K}^{jk}\nabla_k\gamma_{ij} + \frac{1}{2}\bar{K}_{ij}\nabla^j\gamma = \frac{G_4}{12\lambda}\left(\bar{K}_{ij} - \bar{h}_{ij}\bar{K}\right)\nabla^j\alpha, \tag{B.79}$$

$$\dot{\kappa}^{ij} + \dot{\gamma}^{ij}\bar{K} - \bar{h}^{ij}\dot{\kappa} - \dot{\gamma}_{kl}\bar{h}^{ij}\bar{K}^{kl} - \nabla^i\nabla^j n + \bar{h}^{ij}\Delta n + \frac{N}{2}\left(\nabla_k\nabla^j\gamma^{ik} + \nabla_k\nabla^i\gamma^{jk} - \Delta\gamma^{ij} - \nabla^i\nabla^j\gamma\right) = -\frac{G_4}{12\lambda}\bar{N}\nabla^i\nabla^j\alpha, \tag{B.80}$$

and

$$\Delta\dot{\alpha} = \Delta\partial_t\left[\nabla^i\nabla^j\gamma_{ij} - \Delta\gamma + 2\bar{K}_{ij}\kappa^{ij} - 2\kappa\bar{K} - 2\bar{N}^{-1}\left(\Delta n - \nabla_i\gamma^{ij}\nabla_j\bar{N} + \frac{1}{2}\nabla^i\gamma\nabla_i\bar{N}\right)\right] = 0\,. \tag{B.81}$$

The L.H.S. of (B.77)–(B.80) are the same as those in GR and agree with the $\lambda = \xi = 1$ case of [2]. Equation (B.81), however, is very different.

We are interested in perturbations of the modes with wavelengths much shorter than the characteristic scale L of the background, *i.e.*, ω, $p \gg 1/L$, where the space-time is virtually flat. First, (B.81) enforces α to the constant. Thus the R.H.S. of (B.77)–(B.80) are always negligible and thus we have in Fourier space

$$i\omega\gamma^{ij} + 2(\bar{N}\kappa^{ij} + \bar{K}^{ij}n) = 0, \tag{B.82}$$

$$(p_i p_j - p^2\delta_{ij})\gamma^{ij} - 2(\bar{K}\kappa - \bar{K}_{ij}\kappa^{ij}) = 0, \tag{B.83}$$

$$p_j\kappa^{ij} - p^i\kappa - \frac{3}{2}\bar{K}_{jk}\gamma^{jk}p^i + \bar{K}_{jk}p^k\gamma^{ij} + \frac{1}{2}\bar{K}^{ij}p_j\gamma = 0, \tag{B.84}$$

$$i\omega(\delta^{ij}\kappa - \kappa^{ij}) - (p^2\delta^{ij} - p^i p^j)n - i\omega(\bar{K}\gamma^{ij} - \delta^{ij}\bar{K}_{kl}\gamma^{kl}) - \frac{N}{2}(p_k p^j\gamma^{ik} + p_k p^i\gamma^{jk} - p^2\gamma^{ij} - p^i p^j\gamma) = 0. \tag{B.85}$$

Plugging (B.82) into (B.85) and discarding sub-leading terms linear in $\bar{K}_{ij} \sim \mathcal{O}(1/L)$, we simply obtain the fluctuation equation of GR in flat space-time:

$$\frac{\omega^2}{p^2}(\delta^{ij}\kappa - \kappa^{ij}) + i\omega\left(\delta^{ij} - \frac{p^i p^j}{p^2}\right)n + \bar{N}^2\left(\kappa^{ij} + \frac{1}{p^2}\left(p^i p^j\kappa - p_k p^j\kappa^{ik} - p_k p^i\kappa^{jk}\right)\right) = 0. \tag{B.86}$$

Contraction with δ_{ij} and p_j, respectively, yields

$$i\omega\kappa - p^2 n = 0, \tag{B.87}$$

$$p^i\kappa - p_j\kappa^{ij} = 0. \tag{B.88}$$

Using these relations, one can find that (B.86) gives the massless dispersion relation

$$\omega^2 = \bar{N}^2 p^2. \tag{B.89}$$

In this approximation, the rest of the equations, the Hamiltonian and momentum constraints (B.83) and (B.84), are automatically satisfied. In GR, we can set $n = 0$ by using the residual gauge symmetry. Thus we are left with a massless transverse traceless tensor mode, *i.e.*, the usual graviton.

In n-DBI gravity, we do not have liberty to gauge away n. However, we have (B.81) to take into account. Using (B.82), it becomes

$$\begin{aligned}\kappa^{ij}&\left[i\omega\bar{N}\left(\bar{K}_{ij}-\bar{K}\delta_{ij}\right)+\bar{N}^2\left(p^2\delta_{ij}-p_i p_j+\delta_{ij}ip^k\partial_k\ln\bar{N}-2ip_i\partial_j\ln\bar{N}\right)\right]\\&+n\left[-i\omega p^2+\bar{N}\bar{K}_{ij}\left(p^2\delta_{ij}-p_i p_j+ip^k\partial_k\ln\bar{N}\delta_{ij}-2ip_i\partial_j\ln\bar{N}\right)\right]=0.\end{aligned} \tag{B.90}$$

To leading order, this yields

$$\bar{N}^2\kappa^{ij}\left(p^2\delta_{ij}-p_i p_j\right)-in\omega p^2=0 \quad \overset{(B.88)}{\Longrightarrow} \quad n\omega p^2=0. \tag{B.91}$$

Since our approximation is valid only for $\omega, p \gg 1/L$, this implies that $n = 0$ and we are again left with a graviton. To the next order, however, this reads

$$\begin{aligned}\kappa^{ij}&\left[i\omega\bar{N}\left(\bar{K}_{ij}-\bar{K}\delta_{ij}\right)+\bar{N}^2\left(\delta_{ij}ip^k\partial_k\ln\bar{N}-2ip_i\partial_j\ln\bar{N}\right)\right]\\&+n\left[-i\omega p^2+\bar{N}\bar{K}_{ij}\left(p^2\delta_{ij}-p_i p_j\right)\right]=0.\end{aligned} \tag{B.92}$$

Due to (B.87), the κ^{ij} terms contribute. For n to be non-vanishing, we must have the relation

$$\omega \sim \bar{N}p \sim i\left(\bar{N}\bar{K}+\bar{N}\partial\ln\bar{N}\right) \sim \frac{i}{L}. \tag{B.93}$$

However, once again, this cannot be satisfied. Hence there is no scalar mode of the type $(n, \gamma) \sim (n(\omega, p), \gamma(\omega, p))\, e^{i\omega t+ip\cdot x}$ with $|\omega|, |p| \gg 1/L$, including the one with imaginary ω.

B.3.2 Perturbation of the Stückelberg Field

To obtain the equation of motion for the Stückelberg field ϕ, we vary the action (1.140) or (1.141)

$$\delta S \sim e\delta\mathscr{K} \sim e\delta\left(D_\alpha\left(n^\alpha D_\beta n^\beta\right)\right) \sim \left(D_\alpha\left(n^\beta D_\beta e\right)-KD_\alpha e\right)\delta n^\alpha, \tag{B.94}$$

where $e = \left(1+\frac{G_4}{6\lambda}\left({}^{(4)}R+\mathscr{K}\right)\right)^{-\frac{1}{2}}$ and $\mathscr{K} = -2D_\alpha(n^\alpha D_\beta n^\beta)$. Using

$$\delta n^{\alpha} = \frac{\partial^{\alpha}\delta\phi + n^{\alpha}n^{\beta}\partial_{\beta}\delta\phi}{\sqrt{-X}}, \tag{B.95}$$

we find

$$D^{\alpha}\left(\frac{(\partial_{\alpha}+n_{\alpha}n^{\sigma}\partial_{\sigma})(n^{\beta}\partial_{\beta}e) - D_{\beta}n^{\beta}(\partial_{\alpha}+n_{\alpha}n^{\sigma}\partial_{\sigma})e}{\sqrt{-X}}\right) = 0. \tag{B.96}$$

It can be shown that this becomes (3.76) when $n_{\mu} = (-N, 0, 0, 0)$, *i.e.*, $\phi = t$. Note that this equation is invariant under $\phi \to f(\phi)$, as it should be, owing to the fact that $n^{\alpha}(\partial_{\alpha} + n_{\alpha}n^{\sigma}\partial_{\sigma}) = 0$. This also implies that the quantity in the parenthesis is proportional to the space-like vector $n^{\beta}D_{\beta}n_{\alpha}$ tangential to the hypersurface.

Now we consider perturbations $\phi = \bar{\phi} + \varphi$, expanding (B.96) to linear order in φ. We work in unitary gauge, $\bar{\phi} = t$ and $N^i = 0$ and use

$$\delta n^{\alpha} = (0, \bar{N}\partial^{i}\varphi), \qquad \delta\left(\sqrt{-X}\right)^{-1/2} = -\bar{N}\dot{\varphi}, \qquad \delta K = \bar{N}^{-1}\nabla_{i}(\bar{N}^{2}\partial^{i}\varphi), \tag{B.97}$$

as well as

$$\begin{aligned} \delta e &= -\frac{G_4}{12\lambda}\bar{e}^{3}\delta\mathscr{K}, \\ -\frac{1}{2}\delta\mathscr{K} &= \left(2\bar{K} - \bar{N}^{-1}\partial_t\right)\left(\bar{N}^{-1}\nabla_i\left(\bar{N}^{2}\partial^{i}\varphi\right)\right) + \bar{N}\partial^{i}\varphi\partial_{i}\bar{K}. \end{aligned} \tag{B.98}$$

The terms coming from perturbing n^{α} and $1/\sqrt{-X}$ in (B.96) contain at most 3 φ-derivatives, while those from perturbing e contain 6 and 5 φ-derivatives. Thus we only need to consider the latter. It also suffices to keep 3 and 2 δe-derivatives. Then we find

$$\Delta\delta\dot{e} + \bar{K}\bar{N}\Delta\delta e = 0. \tag{B.99}$$

In terms of φ the leading terms (with 6 and 5 derivatives) yield (3.83).

B.3.3 Perturbation of the Stückelberg Field-2nd Order

To find the cubic action of the fluctuation φ in flat space-time (with $q = 1$), we expand the action as

$$S = -\frac{1}{16\pi G_4}\int d^4x\left[\mathscr{K} - \frac{1}{4}\left(\frac{G_N}{6\lambda}\right)\mathscr{K}^2 - \frac{5}{8}\left(\frac{G_N}{6\lambda}\right)^2\mathscr{K}^3 + \cdots\right]. \tag{B.100}$$

The first term is a surface term and does not contribute to the equation of motion. In unitary gauge, the time-like vector takes the form

$$n_0 = -\frac{1+\dot{\varphi}}{\sqrt{(1+\dot{\varphi})^2 - (\partial_i\varphi)^2}}, \qquad n_i = -\frac{\partial_i\varphi}{\sqrt{(1+\dot{\varphi})^2 - (\partial_i\varphi)^2}}. \tag{B.101}$$

This can be expanded as

$$-n_0 = 1 + \frac{1}{2}(\partial_i\varphi)^2 - \dot{\varphi}(\partial_i\varphi)^2 + \mathscr{O}(\varphi^4), \tag{B.102}$$

$$-n_i = \partial_i\varphi - \dot{\varphi}\partial_i\varphi + \dot{\varphi}^2\partial_i\varphi + \frac{1}{2}(\partial_i\varphi)^3 + \mathscr{O}(\varphi^4). \tag{B.103}$$

In the flat background we have

$$\mathscr{K} = -2\left[\partial_0\left(n^0\partial_0 n^0\right) + \partial_i\left(n^i\partial_j n^j\right) + \partial_0\left(n^0\partial_i n^i\right) + \partial_i\left(n^i\partial_0 n^0\right)\right], \tag{B.104}$$

and we find to quadratic order

$$\mathscr{K} = 2\left[\Delta\dot{\varphi} + \partial_0(\dot{\varphi}\Delta\varphi) - \partial_i(\partial_i\varphi\Delta\varphi)\right]. \tag{B.105}$$

Hence the cubic scalar field action is given by

$$S_3 = \frac{1}{192\pi\lambda}\int d^4x\left[\frac{1}{2}(\Delta\dot{\varphi})^2 - (\dot{\varphi}\Delta\ddot{\varphi} - \partial_i\varphi\Delta\partial_i\dot{\varphi})\,\Delta\varphi + \frac{5G_4}{2\lambda}\frac{1}{3!}(\Delta\dot{\varphi})^3\right]. \tag{B.106}$$

B.3.4 A Nonlinear Analysis of the Stückelberg Field

The scalar mode φ obeys the equation of motion (B.96). To determine whether the scalar mode leads to an instability or not, we need to study (B.96) beyond linear order approximation. As we discussed, to linear order, the equation of motion is simply

$$\Delta^2\ddot{\varphi} = 0, \tag{B.107}$$

and the solution is

$$\varphi(t,x) = \varepsilon\left[\varphi_0(x) + \varphi_1(x)t\right], \tag{B.108}$$

where $\varphi_0(x)$ and $\varphi_1(x)$ are arbitrary functions of space. We have included the factor of $\varepsilon \ll 1$ for later convenience. The fully nonlinear solution can in principle be found systematically order by order in ε expansions:

$$\varphi(t,x) = \varepsilon\left[\varphi_0(x) + \varphi_1(x)t\right] + \sum_{n=2}^{\infty}\varepsilon^n\varphi_n(t,x). \tag{B.109}$$

The higher order fluctuations $\varphi_n(t,x)$'s are determined in terms of the initial data $(\varphi_0(x), \varphi_1(x))$ and of order n in powers of (spatial derivatives of) $\varphi_0(x)$ and $\varphi_1(x)$ and polynomial in time t. Using (B.106), for example, the next-to-leading order fluctuation can be found as

$$\varphi_2(t,x) = \frac{1}{2}\varphi_2^{(2)}(x)t^2 + \frac{1}{3!}\varphi_2^{(3)}(x)t^3, \tag{B.110}$$

where

$$\begin{aligned}
\Delta^2\varphi_2^{(2)} &= \Delta\left[2\Delta\varphi_0\Delta\varphi_1 + 2\partial_i\varphi_0\Delta\partial_i\varphi_1 + \Delta\partial_i\varphi_0\partial_i\varphi_1\right] - \left(\Delta\varphi_0\Delta^2\varphi_1 + \Delta\partial_i\varphi_0\Delta\partial_i\varphi_1\right),\\
\Delta^2\varphi_2^{(3)} &= \Delta\left[2(\Delta\varphi_1)^2 + 2(\partial_i\varphi_1)^2 + \partial_i\varphi_1\Delta\partial_i\varphi_1\right] - \left(\Delta^2\varphi_1\Delta\varphi_1 + \partial_i\varphi_1\Delta\partial_i\varphi_1\right).
\end{aligned} \tag{B.111}$$

Clearly, the late time behaviour of the scalar mode requires the knowledge of all order fluctuations. Thus, to see whether the scalar mode yields an instability or not, we need to re-sum the infinite series (B.109). However, this seems to be out of our reach and we will instead resort to an alternative analysis working in the Einstein frame.

B.4 Linear Perturbations in the Einstein Frame

We consider linear perturbations of the scalar fields around flat space-time in the Einstein frame (1.135). The scalar field Lagrangian density reads

$$\begin{aligned}
4\pi G_4\mathscr{L}^E_{\text{scalar}} &= 2\dot\psi^2 + 4\dot\psi(\dot E + \Delta B) - (4n - 2\psi)\Delta\psi\\
&\quad + 4\dot\chi\left(\dot E + \Delta B + 2\dot\psi\right) + 6\dot\chi^2 - 6\chi\Delta\chi - \frac{24\lambda}{G_4}\chi^2.
\end{aligned} \tag{B.112}$$

This is the Einstein frame counterpart of (3.50). The equations of motion are given by

$$\ddot\psi = -\ddot\chi, \tag{B.113}$$
$$\Delta\dot\psi = -\Delta\dot\chi, \tag{B.114}$$
$$\ddot E + \Delta\dot B + \Delta n = -\ddot\chi, \tag{B.115}$$
$$\Delta\psi = 0, \tag{B.116}$$

which clearly reduce to those of GR for constant χ, plus

$$\ddot\chi + \Delta\chi + \frac{1}{3}(\ddot E + \Delta\dot B + 2\ddot\psi) + \frac{4\lambda}{G_4}\chi = 0. \tag{B.117}$$

This is the linearization of (3.87). In the $E = 0$ gauge the general solution can be found as

$$B = B_0(x) + B_1(x)t, \tag{B.118}$$
$$n = -B_1(x), \tag{B.119}$$
$$\psi = 0, \tag{B.120}$$
$$\chi = \chi_0(x), \tag{B.121}$$

with $\chi_0(x)$ related to $B_1(x)$ by

$$\Delta B_1 = -3\Delta\chi_0 - \frac{12\lambda}{G_N}\chi_0. \tag{B.122}$$

Here we have again imposed the boundary condition that all the fields fall off at spatial infinity. Note that $\chi_0(x)$ is essentially $B_1(x)$ which is the degree of freedom responsible for the linear time growth. This suggests the identification $\chi(t, x) \sim \dot{T}(t, x)$, that is, the auxiliary field χ can be regarded as the time derivative of the scalar mode.

References

1. Khoury, J., Miller, G. E. J., & Tolley, A. J. (2011) Spatially covariant theories of a transverse, traceless graviton, part I: Formalism. arXiv:1108.1397 [hep-th].
2. Blas, D., Pujolas, O., & Sibiryakov, S. (2009). On the extra mode and inconsistency of Horava gravity. *JHEP*, *0910*, 029. arxiv.org/abs/0906.3046 [hep-th].

CV-Yuki Sato

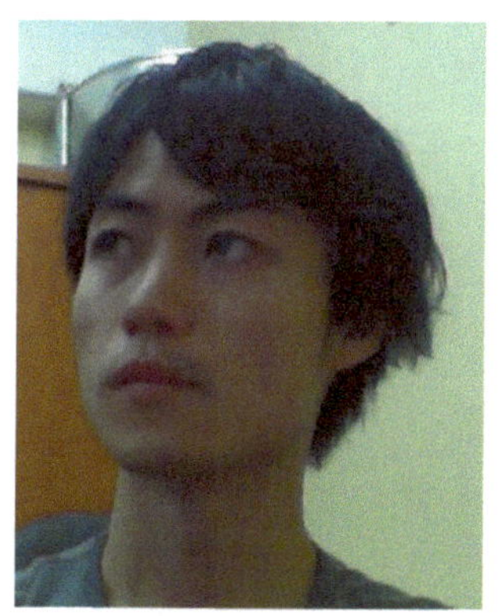

Education

2010–2013	**Ph.D. Physics**, Nagoya University, Japan. Adviser: Prof. Tadakatsu Sakai (in Nagoya University) Adviser: Prof. Jan Ambjørn (in Niels Bohr Institute)
2008–2010	**M.S. Physics**, *Director's prize*, Nagoya University, Japan. Adviser: Prof. Tadakatsu Sakai
2005–2006	Maritime Officer Candidate School, Japan.
Nov. 2003	United Sates Naval Academy, USA.
2001–2005	**B.S. Applied Physics**, *Summa Cum Laude*, National Defense Academy, Japan. Adviser: Prof. Kazuhiko Odaka

Y. Sato, *Space-Time Foliation in Quantum Gravity*, Springer Theses,
DOI: 10.1007/978-4-431-54947-5, © Springer Japan 2014

Work and Research Experiences

Oct. 2013–present	Postdoctoral researcher, Wits University, South Africa.
Apr. 2013–Sep. 2013	Research fellow, KEK, Japan.
Apr. 2010–Mar. 2013	Research assistant, Nagoya University, Japan.
Jun. 2011–Dec. 2012	Visitor, Niels Bohr Institute, Denmark.
Apr. 2005–Oct. 2006	Ensign, Maritime Self Defense Force.

Honors and Awards

1. Director's prize for M.S., 2010.
2. Medal of honor for B.S., 2005.
3. Exchange program with United States Naval Academy, 2003.

Zeitfracht Medien GmbH
Ferdinand-Jühlke-Straße 7
99095 Erfurt, Deutschland
produktsicherheit@kolibri360.de